Lecture Notes in Bioinformatics

Subseries of Lecture Notes in Computer Science

Corrado Priami   Xiaohua Hu   Yi Pan
Tsau Young Lin (Eds.)

Transactions on Computational Systems Biology V

 Springer

Series Editors

Sorin Istrail, Brown University, Providence, RI, USA
Pavel Pevzner, University of California, San Diego, CA, USA
Michael Waterman, University of Southern California, Los Angeles, CA, USA

Editor-in-Chief

Corrado Priami
The Microsoft Research – University of Trento
Centre for Computational and Systems Biology
Piazza Manci, 17, 38050 Povo (TN), Italy
E-mail: priami@msr-unitn.unitn.it

Volume Editors

Xiaohua Hu
Drexel University, College of Information Science and Technology
3141 Chestnut Street, Philadelphia, PA 19104, USA
E-mail: thu@cis.drexel.edu

Yi Pan
Georgia State University, Department of Computer Science
34 Peachtree Street, Atlanta, GA 30302-4110, USA
E-mail: pan@cs.gsu.edu

Tsau Young Lin
San Jose State University, Department of Computer Science
San Jose, CA 95192, USA
E-mail: tylin@cs.sjsu.edu

Library of Congress Control Number: 2006929800

CR Subject Classification (1998): J.3, H.2.8, F.1

LNCS Sublibrary: SL 8 – Bioinformatics

ISSN   1861-2075
ISBN-10   3-540-36048-4 Springer Berlin Heidelberg New York
ISBN-13   978-3-540-36048-3 Springer Berlin Heidelberg New York

Springer is a part of Springer Science+Business Media

springer.com

© Springer-Verlag Berlin Heidelberg 2006

Typesetting: Camera-ready by author, data conversion by Scientific Publishing Services, Chennai, India
Printed on acid-free paper   SPIN: 11790105   06/3142   5 4 3 2 1 0

Preface

This issue of *Transactions on Computational Systems Biology* contains a selection of papers presented initially at the 2005 IEEE International Conference on Granular Computing held in Beijing, July 25–27, and a few invited papers. Papers included in this special issue are devoted to various aspects of computational methods, algorithms, and techniques in bioinformatics such as gene expression analysis, biomedical literature mining and natural language processing, protein structure prediction, biological database management and biomedical information retrieval.

Z. Huang, Y. Li and X. Hu present a novel SVM-based method to predict anti-parallel structure from sequence data.

C.H. Liu, I.-J. Chiang, J.-M. Wong, H.-C. Tsai and T.Y. Lin introduce a novel model of concept representation called Latent Semantic Networks using a multilevel geometric structure.

B. Jin and Y.-Q. Zhang propose a new system to evolve the structures of granular kernel trees (GKTs) in the case that we lack knowledge to predefine kernel trees. The new granular kernel tree structure evolving system is used for cyclooxygenase-2 inhibitor activity comparison.

M.K. Ng, S.-Q. Zhang, W.-K. Ching and T. Akutsu study a control model for gene intervention in a genetic regulatory network. At each time step, a finite number of controls are allowed to drive to some target states (i.e., some specific genes are on, and some specific genes are off) of a genetic network.

Z. Peng, Y. Shi and B. Zhai discuss how to manage a large amount of complex biological data by an object deputy database system which can provide rich semantics and enough flexibility. In their system, the flexible inheritance avoids a lot of data redundancy.

R. Satre, H. Sovik, T. Amble and Y. Tsuruoka address the natural language understanding in molecular biology literature. Their prototype system GeneTUC is capable of doing deep reasoning, such as anaphora resolution and question answering, which is not a part of the state-of-the-art parsers.

H.-C. Wang, Y.-S. Lee and T.-H. Huang describe a novel approach to combine microarray data and literature to find the relations among genes. Unlike other techniques, this method not only reduces the comparison complexity but also reveals more mutual interactions among genes.

H.-H. Yu, V.S. Tseng and J.-H. Chuang propose a multi-information-based methodology to score genes based on the microarray expressions. The concept of multi-information here is to combine different scoring functions in different tiers for analyzing gene expressions. The proposed methods can rank the genes according to the degree of relevance to the targeted diseases so as to form a precise prediction base.

X. Zhou, X. Hu, G. Li, X. Lin and X. Zhang explore the use of term relations in information retrieval for precision-focused biomedical literature search. A relation is defined as a pair of two terms which are semantically and syntactically related to each other. Unlike the traditional "bag-of-word" model for documents, their model represents a document by a set of sense-disambiguated terms and their binary relations. A prototyped IR system supporting relation-based search is then built for Medline abstract searches. The experiment shows the expressiveness of relations for the representation of information needs, especially in the area of biomedical literature full of various biological relations.

We would like to thank the authors for contributing their research work to the special issue as well as the Editor-in-Chief of the LNCS Transaction on Computational Systems Biology, Prof. Priami.

The editors of the special issue:
Xiaohua Hu, Drexel University
Yi Pan, Georgia State University
T.Y. Lin, San Jose State University

LNCS Transactions on
Computational Systems Biology –
Editorial Board

Table of Contents

Anti-parallel Coiled Coils Structure Prediction by Support Vector
Machine Classification
  Zhong Huang, Yun Li, Xiaohua Hu 1

A Complex Bio-networks of the Function Profile of Genes
  Charles C.H. Liu, I-Jen Chiang, Jau-Min Wong,
  Ginni Hsiang-Chun Tsai, Tsau Young ('T.Y.') Lin 9

Evolutionary Construction of Granular Kernel Trees for
Cyclooxygenase-2 Inhibitor Activity Comparison
  Bo Jin, Yan-Qing Zhang 25

A Control Model for Markovian Genetic Regulatory Networks
  Michael K. Ng, Shu-Qin Zhang, Wai-Ki Ching, Tatsuya Akutsu 36

Realization of Biological Data Management by Object Deputy
Database System
  Zhiyong Peng, Yuan Shi, Boxuan Zhai 49

GeneTUC, GENIA and Google: Natural Language Understanding
in Molecular Biology Literature
  Rune Sætre, Harald Søvik, Tore Amble, Yoshimasa Tsuruoka 68

Gene Relation Finding Through Mining Microarray Data
and Literature
  Hei-Chia Wang, Yi-Shiun Lee, Tian-Hsiang Huang 83

A Multi-information Based Gene Scoring Method for Analysis of Gene
Expression Data
  Hsieh-Hui Yu, Vincent S. Tseng, Jiin-Haur Chuang 97

Relation-Based Document Retrieval for Biomedical IR
  Xiaohua Zhou, Xiaohua Hu, Guangren Li, Xia Lin,
  Xiaodan Zhang 112

Author Index 129

Anti-parallel Coiled Coils Structure Prediction by Support Vector Machine Classification

Zhong Huang, Yun Li, and Xiaohua Hu

College of Information Science and Technology, Drexel University,
3141 Chestnut Street, Philadelphia, PA, USA, 19104
thu@cis.drexel.edu

Abstract. Coiled coils is an important 3-D protein structure with two or more stranded alpha-helical motif wounded around to form a "knobs-into-holes" structure. In this paper we propose an SVM classification approach to predict the anti-parallel coiled coils structure based on the primary amino acid sequence. The training dataset for the machine learning are collected from SOCKET database which is a SOCKET algorithm predicted coiled coils database. Total 41 sequences of at least two heptad repeats of the anti-parallel coiled coils motif are extracted from 12 proteins as the positive datasets. Total 37 of non coiled coils sequences and parallel coiled coils motif are extracted from 5 proteins as negative datasets. The normalized positional weight matrix on each heptad register a, b, c, d, e, f and g is from SOCKET database and is used to generate the positional weight on each entry. We performed SVM classification using the cross-validated datasets as training and testing groups. Our result shows 73% accuracy on the prediction of anti-parallel coiled coils based on the cross-validated data. The result suggests a useful approach of using SVM to classify the anti-parallel coiled coils based on the primary amino acid sequence.

Keywords: coiled coil, SOCKET algorithm, SVM, protein sequence data.

1 Introduction

Coiled coils structure was first introduced by Crick in 1953 in which he postulated a hallmark structure of "knobs-into-holes" formed by wounded strands of alpha-helices [2]. The coiled coils structure is characterized by a heptad repeats of amino acids $(a\text{-}b\text{-}c\text{-}d\text{-}e\text{-}f\text{-}g)n$. Positions a and d in one chain are occupied by apolar hydrophobic amino acids to form the core packing structure with the same positions in partner chain (see figure 1). The coiled coils structure is further stabilized by side chain electrostatic interaction of $e\text{-}g$ between two chains which generally occupied by polar charged amino acids. Recently it has been shown that intrachain interactions between heptad residues also contribute to the stability of the coiled coils structure [6] [12].

Due to its well characterized structure the coiled coils has long been a spotlight of the protein design and prediction study. However, the structure of coiled coils is of great diversity in terms of its interchain orientation and oligomer status. The coiled coils structure can be formed between two, three, four or even five chains and the

C. Priami et al. (Eds.): Trans. on Comput. Syst. Biol. V, LNBI 4070, pp. 1–8, 2006.
© Springer-Verlag Berlin Heidelberg 2006

orientation of each chain can be the same (parallel) or different (antiparallel). The core packing registers *a* and *d* are important for determining the number of strands while e-g interaction seems to be important in choosing the helices partners [3] [4] [7] [11]. Therefore the primary sequence of the heptad repeats may be one of the determining factors on the specificity of the coiled coils but much is still poorly understood so far [12].

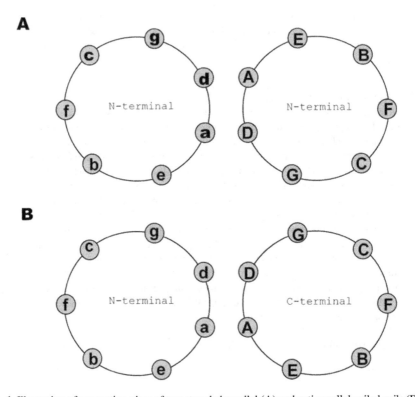

Fig. 1. Illustration of transaction view of two stranded parallel (A) and anti-parallel coiled coils (B)

Two categories of algorithms have been proposed to predict the coiled coils structure based on either the primary amino acid sequence or the atomic 3-D coordinate information. The first category includes COILS [5], PAIRCOIL [1] and MULTICOIL [10]. These algorithms compare the sequence of the target protein with the amino acid sequence database of known two or three stranded parallel coiled coils and give the score of probability. However, none of them are suitable for predicting the antiparallel coiled coils. The second category of prediction algorithm utilizes the 3-D coordinate information of the polypeptide to predict and define the beginning and ending of coiled coils based on the database of known 3-D structure of coiled coils. SOCKET [9] and TWISTER [8] are two algorithms in this category. The SOCKET algorithm focuses on the core packing structure of the "knobs-into-holes" which is formed by interchain *a-d* interactions. The TWISTER algorithm is designed to identify not only the canonical coiled coils but also the special coiled coils with discontinuous heptad repeats interrupted by stutters and skips. Both algorithms take the 3-D atomic coordinate PDB file and DSSP file as input and are able to predict two or three stranded parallel and anti-parallel coiled coils. Comparing with the first category of algorithms, using 3-D coordinate as input may seem to be a better choice as the SOCKET algorithm attempts to identify the core packing structure of coiled coils based on the experimentally determined protein 3-D structure. The disadvantage of utilizing SOCKET to predict coiled coils structure is that it requires atomic coordinate information of the polypeptide which may not be readily available. However, with the rapid expanding of the PDB database which currently already collects over 34,000 structures determined by x-ray crystallization and NMR spectroscopy [13], such algorithms can be adopted more widely to predict coiled coils structure.

In this paper we used the machine learning supporting vector machine (SVM) approach to discriminate the anti-parallel coiled coils structure based on the primary amino acid sequence using the normalized amino acid profile of heptad repeat generated by SOCKET [9]. We selected anti-parallel coiled coils proteins with at least two full set of heptad repeats from SOCKET database such that each vector has the same number of features for SVM training and testing. Our preliminary results suggest that SVM is a valuable tool to predict the anti-parallel coiled coils based on the amino acid profile originally determined by atomic 3-D coordinate of the protein.

2   Methods, Results and Discussion

We selected total 78 anti-parallel coiled coils from SOCKET database which currently lists a total of 134 entries [9] based on PDB release #89. The PDB files of 78 anti-parallel coiled coils and 8 parallel coiled coils were downloaded using a perl script available from PDB ftp site. The PDB files from both the anti-parallel and parallel coiled coils were submitted to SOCKET server (http://www.biols.susx.ac.uk/ Biochem/Woolfson/html/coiledcoils/socket/server.html) to identify the specific heptad repeats registers. Only the long anti-parallel coiled coils with at least two full sets of the heptad repeats were selected. This is mainly based on our assumption that long coiled coils are more structurally stable and may include more positional information which could contribute to the stability and specificity of coiled coils. Considering the relatively small number of entries currently available, we allow multiple contributions of coiled coils structure from the same protein. Because the SVM only accepts vectors with the same number of features, we chose 2 heptad repeats of total 14 amino acids from each entry. In the case of heptad repeats are more than 2, we allow partial overlap of the heptad repeats assuming each partially overlapped heptad repeats is an independent vector for SVM to avoid loss of any given heptad repeat. Total 41 sequences of at least two heptad repeats of the anti-parallel coiled coils motif are extracted from 12 proteins as the positive datasets (Table 1). Total 37 of non coiled coils sequences and parallel coiled coils motif are extracted from 5 proteins as negative datasets.

We used the normalized frequencies of occurrence at each heptad positions for long anti-parallel coiled coils from Walshaw et al [9] to convert the amino acid of each heptad sequence into amino acid usage frequencies (shown in Table 2).

Table 1. Positive datasets of proteins with anti-parallel coiled coil structures

PDB ID	Protein name
5eau	5-Epi-Aristolochene Synthase
2spc	Spectrin
2ktq	Large Fragment Of DNA Polymerase I
2fha	Human H Chain Ferritin
1ser	seryl-tRNA synthetase complexed with tRNA(Ser)
1ecr	Replication Terminator Protein (Tus) Complexed With DNA
1ecm	Chorismate Mutase
1cnt	Ciliary Neurotrophic Factor
1cii	Colicin Ia
1aqt	F1F0-ATP Synthase
1ab4	59Kda Fragment Of Gyrase A
1a36	Human DNA Topoisomerase I

Table 2. Normalized frequencies of occurrence at each position of the heptad sequence for two stranded long anti-parallel coiled coils from Walshaw et al [9]

Corner	a	b	c	d	e	f	g
A	1.42	1.48	1.28	1.55	0.69	1.04	1.98
C	0	0	0	0.59	0	1.19	0
D	0.09	1.06	1.71	0.56	1.37	1.75	0.25
E	0.54	1.32	1.75	1.31	1.45	1.76	1.23
F	0.84	0.51	0.85	0.96	0.8	0.96	0.64
G	0	0.72	0.81	0.07	0.48	0.38	0.19
H	1.09	1.56	0.31	0.44	0.88	1.47	3.79
I	2.87	0.48	0.24	1.6	1.59	0.79	1.12
K	0.33	2.24	1.52	0.41	1.22	1.55	0.66
L	2.96	1.04	1.33	3.64	1.31	0.35	1.11
M	1.45	1.18	0.59	1.65	0.83	0.83	1.93
N	0.88	1.58	0.94	0.22	0.44	1.78	0.59
P	0	0	0.28	0	0.27	0	0
Q	0.86	1.23	1.75	0.74	1.99	1.16	2.3
R	0.66	1.36	1.48	0.47	2.55	1.78	1.14
S	1.03	0.78	0.58	0.28	0.37	0.83	0.83
T	0.61	1.23	0.73	0.61	0.58	1.39	1.15
V	0.51	0.21	0.53	0.37	0.61	0.41	0.99
W	0	0.56	2.24	0.31	0.53	0.53	0.53
Y	1.23	0.66	0.44	1.69	0.82	0.41	0.2

Table 3. The testing results of SVM using cross-validation approach

Corner	Class Label	Prediction by SVM
1a36-1	1	1
1a36-2	1	-1
1ab4-1	1	1
1ab4-2	1	-1
1aqt-1	1	-1
1cii-1	1	1
1cii-2	1	1
1cii-3	1	1
1cii-4	1	-1
1cii-5	1	1
1cii-6	1	-1
1cii-7	1	-1
1cii-8	1	-1
1cnt-1	1	1
1cnt-2	1	1
1cnt-1	1	1
1cnt-2	1	1
1cnt-3	1	1
1cnt-4	1	1
1cnt-5	1	1
1ecm-1	1	1
1ecm-2	1	1
1ecm-3	1	1
1ecm-4	1	1
1ecr-1	1	1
1ecr-2	1	1
1ser-1	1	1
1ser-2	1	-1
2fha-1	1	1
2fha-2	1	-1
2fha-3	1	-1
2fha-4	1	1
2fha-5	1	1
2ktq	1	1
2spc-1	1	1
2spc-2	1	1
2spc-3	1	1
2spc-4	1	1
2spc-5	1	1

Table 3. *(Continued)*

2spc-6	1	1
5eau	1	1
m6a-1	-1	-1
m6a-2	-1	-1
m6a-3	-1	-1
m6a-4	-1	-1
m6a-5	-1	-1
m6a-6	-1	1
m6a-7	-1	-1
m6a-8	-1	-1
m6a-9	-1	1
m6a-10	-1	-1
m6a-11	-1	1
m6a-12	-1	-1
m6a-13	-1	1
m6a-14	-1	-1
m6a-15	-1	-1
1a93-1	-1	-1
1a93-2	-1	-1
1a93-3	-1	1
1a93-4	-1	1
1fos-1	-1	-1
1fos-2	-1	-1
1fos-3	-1	1
1fos-4	-1	-1
1fos-5	-1	-1
1fos-6	-1	-1
1fos-7	-1	1
1fos-8	-1	1
1fos-9	-1	-1
1fos-10	-1	-1
mbplike-1	-1	1
mbplike-2	-1	-1
mbplike-3	-1	-1
mbplike-4	-1	-1
mbplike-5	-1	-1
mbplike-6	-1	1
mbplike-7	-1	-1
MBP	-1	-1

Two programs written in perl and java were used to convert the list of 14 amino acid sequence into amino acid usage frequency at each position and to generate label file for SVM classification. Due to the small number of samples, we adopted cross-validation approach for SVM training and testing. The total 78 datasets are separated into two groups, with 77 datasets for training and 1 dataset for testing in each cycle. Web interface of SVM (Gist version 2.0.5, http://svm.sdsc.edu) was used in dataset training and testing. The testing results are shown in Table 3.

From total 78 datasets including 41 positive and 37 negative datasets, we identified 31 true positive and 26 true negative respectively using SVM classification. Our results show that the average accuracy for the testing is 73% (summarized in Table 4).

Table 4. Average Accuracy Result

Total Dataset	78
Positive	41
Negative	37
Testing Result	
False Positive	11
False Negative	10
True Positive	31
True Negative	26
Average Accuracy	73%

We demonstrated that SVM classification algorithm can be used to discriminate anti-parallel coiled coils structure from non-parallel coiled coils including the parallel coiled coils structure. Each protein sequence with two heptad repeats was taken as input vector with 14 dimensional features and subjected to SVM training and testing. All 14 features of each vector have same semantics representing the amino acid usage frequencies. Our result indicated that SVM learning algorithm can discriminate two classes using hyperplane with maximum margin between vectors of two classes with relatively high accuracy.

Walshaw et al calculated normalized amino acid frequencies of occurrence at both the parallel and anti-parallel coiled coils and found that anti-parallel coiled coils tends to have broader amino acid usages at each heptad position compared with its parallel counterpart [9]. This characteristic of amino acid profile associated specifically with anti-parallel coiled coils also prompted us to select higher dimensional features which include two repeats of heptad, instead of focusing only on hallmark feature of "knob-into-hole" structure at a/d positions.

3 Conclusion and Future Plan

In this paper we presented a SVM approach for classification of anti-parallel coiled coils structure based on primary sequences. The classification results indicate that the support vector machine learning algorithm is a useful tool in classifying the anti-parallel coiled coils structure, which by far has no direct algorithm to predict its structure based on its primary sequence. However, more datasets are needed to further validate the approach. With the rapid growing of the PDB database, the SOCKET database has been growing accordingly. We expect the future release of SOCKET database will be helpful in gathering more positive datasets and help to improve the SVM prediction accuracy.

Acknowledgement. This work is supported partially by the NSF Career grant IIS 0448023 and NSF 0514679 and PA Dept of Health Tobacco Formula Grants.

References

1. Berger, B. (1995). "Algorithms for protein structural motif recognition." J Comput Biol 2(1): 125-38.
2. Crick, F. H. C. (1953). "The packing of alpha-helices: simple coiled-coils." Acta. Crystallog. 6: 689-697.
3. Harbury, P. B., T. Zhang, et al. (1993). "A switch between two-, three-, and four-stranded coiled coils in GCN4 leucine zipper mutants." Science 262(5138): 1401-7.
4. Kohn, W. D., C. M. Kay, et al. (1998). "Orientation, positional, additivity, and oligomerization-state effects of interhelical ion pairs in alpha-helical coiled-coils." J Mol Biol 283(5): 993-1012.
5. Lupas, A., M. Van Dyke, et al. (1991). "Predicting coiled coils from protein sequences." Science 252(5010): 1162-4.
6. Oakley, M. G. and J. J. Hollenbeck (2001). "The design of antiparallel coiled coils." Curr Opin Struct Biol 11(4): 450-7.
7. O'Shea, E. K., R. Rutkowski, et al. (1992). "Mechanism of specificity in the Fos-Jun oncoprotein heterodimer." Cell 68(4): 699-708.
8. Strelkov, S. V. and P. Burkhard (2002). "Analysis of alpha-helical coiled coils with the program TWISTER reveals a structural mechanism for stutter compensation." J Struct Biol 137(1-2): 54-64.
9. Walshaw, J. and D. N. Woolfson (2001). "Socket: a program for identifying and analysing coiled-coil motifs within protein structures." J Mol Biol 307(5): 1427-50
10. Wolf, E., P. S. Kim, et al. (1997). "MultiCoil: a program for predicting two- and three-stranded coiled coils." Protein Sci 6(6): 1179-89
11. Woolfson, D. N. and T. Alber (1995). "Predicting oligomerization states of coiled coils." Protein Sci 4(8): 1596-607
12. Yu, Y. B. (2002). "Coiled-coils: stability, specificity, and drug delivery potential." Adv Drug Deliv Rev 54(8): 1113-29.
13. Berman H.M., Westbrook J. et al (2000). "The Protein Data Bank." Nucleic Acids Res 28 235-242.

A Complex Bio-networks of the Function Profile of Genes

Charles C. H. Liu[1,3], I-Jen Chiang[2,3,*], Jau-Min Wong[3], Ginni Hsiang-Chun Tsai[3], and Tsau Young ('T. Y.') Lin[4]

[1] Department of Surgery, Cathay Medical Center, Taipei, Taiwan 106
chliu@ntu.edu.tw
[2] Graduate Institute of Medical Informatic, Taipei Medical University,Taipei, Taiwan 110
ijchiang@tmu.edu.tw
[3] Graduate Institue of Medical Engineering, National Taipei University, Taipei, Taiwan 100
[4] Department of Computer Science, San Jose State University, San Jose, CA 95192-0249
tylin@cs.sjsu.edu

Abstract. This paper presents a novel model of concept representation using a multilevel geometric structure, which is called *Latent Semantic Networks*. Given a set of documents, the associations among frequently co-occurring terms in any of the documents define naturally a geometric complex, which can then be decomposed into connected components at various levels.

This hierarchical model of knowledge representation was validated in the functional profiling of genes. Our approach excelled the traditional approach of vector-based document clustering by the geometrical forms of frequent itemsets generated by the association rules. The biological profiling of genes were a complex of concepts, which could be decomposed into primitive concepts, based on which the relevant literature could be clustered in adequate "resolution" of contexts. The hierarchical representation could be validated with tree-based biomedical ontological frameworks, which had been applied for years, and been recently enriched by the online availability of *Unified Medical Language System (UMLS)* and *Gene Ontology (GO)*.

Demonstration of the model and the clustering would be performed on the relevant *GeneRIF (References into Function)* document set of NOD2 gene. Our geometrical model is suitable for representation of bio-logical information, where hierarchical concepts in different complexity could be explored interactively according to the context of application and the various needs of the researchers. An online clustering search engine for use on general purpose and for biomedical use, managing the search results from Google or from PubMed, are constructed based on the methodology (http://ginni.bme.ntu.edu.tw). The hierarchical presentation of clustering results and the interactive graphical display of the contents of each cluster shows the merits of our approach.

1 Introduction

One of the urgent need of bioinformatics in the post-genomic era is to find "biological themes" or "topics" between genes or gene products, in order to "drink from the fire hose" from vast amounts of literature and ex-periment results.

---

* Corresponding author.

C. Priami et al. (Eds.): Trans. on Comput. Syst. Biol. V, LNBI 4070, pp. 9–24, 2006.
© Springer-Verlag Berlin Heidelberg 2006

One approach of theme finding is to derive knowledge directly without translation by another knowledge source, e.g. a vocabulary system. One of the early successful approaches is direct mining from the source literature. The relationships between genes are constructed by probabilistic modes, such as Bayesian Networks. The most clinically yielding is the PubGene project [4]. However, the interpretation of the results is often qualitative, selectively on some local findings in large graph models. The lack of overall picture is partly due to the exploration of individual genes without preliminary grouping of some closely correlated genes. The result relied on the quality of documents collected as "relevant" to the target genes [8].

Subsequent researches to find "molecular pathway" in raw documents is vigorous use of natural language processing techniques. One of the efforts with a long history of literature mining in other medical domain is the GENIE project, evolved from MEDLEE works [2]. Finely tuned rule-based term tagging and processing improve the efficiency, but the rule sets or knowledge sources they constructed cannot be reused by other applications or be validated by others. Besides, the system is too large for personal document browsing.

The other approaches use external knowledge system, such as keyword hierarchy, to group the raw gene information to more biologically under-standable "themes". The early works are well reviewed by Shatkay in the analysis of microarray data [8]. MedMesh is more recent work addressing on the MeSH systems (Medical Subject Heading) of UMLS (Unified Medical Language System), but much raw document processing is used and the approach was relatively in a "black box" [6]. After the advent of Gene Ontology (GO) system, more tools were developed to apply the ontological framework to impose domain knowledge on analysis of raw data, which were listed under the section of "GO tools" in the official site of the GO Consortium [1].

From the medical point of view, current application of MeSH or GO is still in a very primitive developing stage. One of the main reason lies on the nature of tree-based ontological system. For example, GO divides the functional profiles into three branches from the root – the function do-main, the process domain, and the anatomical domain. The first two do-mains are closely associated in many applications. The third domain is also dependent on the first two "function" domains. In addition, the amount of annotations of genes to the three domains is also unbalanced.

Our research addresses on the limitation of functional analysis of genes by the traditional approaches, and proposed a new geometric model. In what follows, we start by reviewing related work on the models of the relationships between gene and gene products clustering in section 2. The concepts and definitions of *latent semantic networks* based on geometric forms for the frequent itemsets generated by association rules are given in section 3. The clustering results for clustering of the functioning profile of a gene are described in Section 4 and Section 5; followed by the conclusion.

2 Related Work

Detecting knowledge based on the co-occurrence of terms or concepts is one of the basic mechanism of document clustering, and was initially proposed to cluster genes into biologically meaningful groups [4]. However, the characteristics of the "groups" could not be explained by the co-occurrence alone. An approach of getting the biological "meaning" was by annotation with associated MeSH and GO terms, which were both tree-based. Our work approaches the "meaning" problem by proposing a new geometric model of clustering in order to more adequately present the network nature of the functioning profiles of genes.

After Girvan and Newman's work of "community structure" in social and biological networks [3], the nature of graph structure inherent in a co-occurrence network began to be explored. Wilkinson et al. [6] picked sets of genes correlated to user-selected keywords by partitioning the components of gene co-occurrence networks functionally correlated "communities". Wren et al. [9] studied the connections in the gene network to rank the "cohesiveness" of co-occurring genes, diseases, and chemical compounds.

The current published genetic analyses based on "community networks" were calculated based on geometrical measurement in the Euclidean space, which we considered is a fundamental limitation of statistical calculation in document or concept clustering. The clustering of distance measurements between sets of more primitive concepts to form higher hierarchy of concept groups is more applicable in topological spaces than in Euclidean spaces. We proposed a topologically based network more suitable for gene analysis.

Based on the network model, we constructed an online "clustering search" engine, which received PubMed and Google queries results, selected the significant concepts, and provided hierarchical and graphical views of the associations between concepts on the fly. The details could be explored by the readers according to their needs interactively. By the example of recently vigorously explored NOD2 gene, which is closely associated with the immunity and inflammatory responses of the inflammatory bowel disease (IBD), we demonstrated the practical applications of our model and clustering methodology in functional profiling of genes. Automatic generation of graphical relationships between concepts from the user specified scope also provoked new insights into the structure of knowledge and was beneficial for improvement of current tree-based vocabulary systems.

3   Geometric Representation of Concept

Term-term inter-relationships that are denoted by their co-occurred associations can automatically model and extract the concepts from a collection of documents. These concepts organize a multilevel and homogenous hierarchy called a *Latent Semantic Network*. The most natural way to represent a latent semantic network is expressed by using the geometric and topologic notations, which can capture the totality of thoughts expressed in this collection of documents; and a "simple component" (which is a *r-connected component*) of a level of hierarchy represents some concept inside this collection.

3.1   Combinatorial Geometry

Let us introduce and define some combinatorial topological concepts. The central idea is n-simplex.

Definition 1. *A n-simplex is a set of independent abstract vertices* $[v_0, \ldots, v_{n+1}]$.

Geometrically 0-simplex is a vertex, 1-simplex an edge (a vertex pair), 2-simplex a triangle, 3-simplex a tetrahedron. A n-simplex is the $n + 1$ dimensional analog. This is the smallest convex set in a Euclidean space R^{n+1} containing $n + 1$ points $v_0 \ldots,$ v_{n+1} that do not lie in a hyperplane of dimension less than n. For example, there is the standard n-simplex

$$\delta^n = \{(t_0, t_1, \ldots t_{n+1}) \in R^{n+1} \mid \sum_i t_i = 1, t_i \geq 0\}$$

Definition 2. *A face of a n-simplex* $[v_0, \ldots, v_{n+1}]$ *is a r-simplex* $[v_{j_0}, \ldots, v_{j_{r+1}}]$ *whose vertices is a subset* $\{v_0, \ldots, v_{n+1}\}$ *with cardinality* $r + 1$.

Definition 3. *A complex is a finite set of simplices that satisfies the following two conditions:*

- *Any face of a simplex from a complex is also in this complex.*
- *The intersection of any two simplices from a complex is either empty or is a face for both of them.*

The vertices of the complex v_0, v_1, $\cdots$, v_n *is the union of all vertices of those simplices.* [7]

Definition 4. *A hereditary* n *simplex, or abbreviated to be n-H-simplex is a special complex of n dimensions that consists of one n-simplex and all its faces.*

Definition 5. *A* (n, r)*-skeleton (denoted by* S_r^n*) of n-complex is a n-complex whose k-faces($k \leq r$) are removed.*

Definition 6. *For any non-empty two simplices A, B are said to be* r-connected *if there exits a sequence of k-simplices* $A = S_0, S_1, \ldots, S_m = B$ *such that* S_j *and* S_{j+1} *has an h-common face for* $j = 0, 1, 2, \ldots, m - 1$*; where* $r \leq h \leq k \leq n$.

Definition 7. *The maximal r-connected subcomplex is called a* r-connected *component. Note For a r-connected component implies there does not exist any r-connected component that is the superset of it.*

3.2   Simple Concept Geometric Structure

In our application each vertex is a key term, so a simplex defines a set of key terms in a collection of documents. Hence, we believe a simplex represents a primitive concept in the collection. For example, the 1-simplex [Wall, Street] represents a primitive concept in financial business. The 0-simplex [Network] might represent many different concepts, however, while it is combined with some other terms would denote latent semantic concepts, such as, these 1-simplices [Computer, Network], [Traffic, Network], [Neural, Network], [Comunication, Network], and so on, demonstrate distinct concepts and identify more obvious semantic than 0-simplex. Of course, the 1-simplex [Neural, Network] is not conspicuous than the 2-simplices [Artifical Neural Network] and [Biology, Neural, Network].

A collection of documents most likely consists of several distinct primitive concepts. Such a collection of primitive concepts is combinatorial a complex.

An idea (in the forms of complex of keywords) may consist of a lot of primitive concepts (in the form of simplices) that are embedded in a document collection. Some primitive concepts may share a common primitive concept, some may not. This situation may be captured by a combinatorial complex of key terms: An idea in the forms of a complex of keywords may consist of a lot of primitive concepts in the form of simplices. Some primitive concepts (simplices) may share a common concept (a common face), some may not.

Example 1. *In Figure 4, we have an idea that consist of twelve terms that organized in the forms of 3-complex. Two $Simplex(a, b, c, d)$ and $Simplex(w, x, y, z)$ are two maximal H-simplices with the highest rank 3. Considering $(3, 1)$-skeleton, S_1^3, by removing*

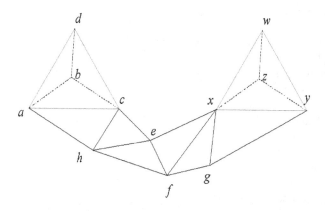

Fig. 1. A complex with twelve vetrics

all 0-simplices, all the other simplices in it can be listed as follows.

- Simplex(a, b, c, d) *and its ten subsimplices:*
  - Simplex(a, b, c)
  - Simplex(a, b, d)
  - Simplex(a, c, d)
  - Simplex(b, c, d)
  - Simplex(a, b)
  - Simplex(a, c)
  - Simplex(b, c)
  - Simplex(a, d)
  - Simplex(b, d)
  - Simplex(c, d)
- Simplex(a, c, h) *and its three subsimplices:*
  - Simplex(a, c)
  - Simplex(a, h)
  - Simplex(c, h)

– Simplex(c, h, e) *and its three subsimplices:*
  - Simplex(c, h)
  - Simplex(h, e)
  - Simplex(c, e)
– Simplex(e, h, f) *and its three subsimplices:*
  - Simplex(e, h)
  - Simplex(h, f)
  - Simplex(e, f)
– Simplex(e, f, x) *and its three subsimplices:*
  - Simplex(e, f)
  - Simplex(e, x)
  - Simplex(f, x)
– Simplex(f, g, x) *and its three subsimplices:*
  - Simplex(f, g)
  - Simplex(g, x)
  - Simplex(f, x)
– Simplex(g, x, y) *and its three subsimplices:*
  - Simplex(g, x)
  - Simplex(g, y)
  - Simplex(x, y)
– Simplex(w, x, y, z) *and its ten subsimplices:*
  - Simplex(w, x, y)
  - Simplex(w, x, z)
  - Simplex(w, y, z)
  - Simplex(x, y, z)
  - Simplex(w, x)
  - Simplex(w, y)
  - Simplex(w, z)
  - Simplex(x, y)
  - Simplex(x, z)
  - Simplex(y, z)

Simplex(a, c), Simplex(c, h), Simplex(h, e), Simplex(e, f), Simplex(f, x), Simplex (g, x), *and* Simplex(x, y) *are common faces that generate a connected path from* Simplex(a, b, c, d) *to* Simplex(w, x, y, z). *There exists a single maximal connected component. Furthermore, considering the* $(3, 2)$*-skeleton,* S_2^3, *by eliminating all* 0*-simplices and* 1*-simplices, all the remainder simplices of it are as follows.*

– Simplex(a, b, c, d) *and its four subsimplices:*
  - Simplex(a, b, c)
  - Simplex(a, b, d)
  - Simplex(a, c, d)
  - Simplex(b, c, d)
– Simplex(a, c, h)
– Simplex(c, h, e)
– Simplex(e, h, f)
– Simplex(e, f, x)

- Simplex(f, g, x)
- Simplex(g, x, y)
- Simplex(w, x, y, z) *and its four subsimplices:*
  - Simplex(w, x, y)
  - Simplex(w, x, z)
  - Simplex(w, y, z)
  - Simplex(x, y, z)

There does not exist any common faces between any two simplices, so that eight maximal connected components are in S_2^3. So does S_3^3, there are only two maximal connected components in it because the maximum rank of simplices in it is 3.

A maximal connected component of a skeleton represents a complex of association rules, i.e., a set of concepts. If a maximal connected component of a skeleton contains only one simplex, this component is said to organize a primitive concept.

Definition 8. *A maximal connected component is said to be* independent *if it is composed of a single simplex, i.e., there is no common face between two maximal connected components.*

3.3 Issues

From a collection of documents, a complex of association rules can be generated. A skeleton of a complex is closed, because all subcomplexes of a complex are also in the skeleton according to subsimplices in each composite simplex of a complex in a skeleton are also included in the simplex, which satisfies the *apriori* property. As seen in Example 1, all connected components in S_k^n are contained in S_r^n, where $k \geq r$. Based on that, the goal of this paper is to establish the following belief.

Claim. A maximal independent connected component of a skeleton represents a primitive *concept* in this collection of documents.

Example 2. *Given a skeleton, S_1^2, of association rules depicted in Figure 2, it is a 2-complex composed of the term set $V = \{t_A, t_B, t_C\}$ in a collection of documents. In the skeleton, all 0-simplices are neglect, i.e., the terms depicted in dash lines. The simplex set $S = \{\text{Simplex}_1, \text{Simplex}_2, \text{Simplex}_3, \text{Simplex}_4\}$ (Simplex$_1$ is a 2-simplex and Simplex$_2$, Simplex$_3$ as well as Simplex$_4$ are 1-simplices) represents generated frequent item-sets from V, and $W = \{w_{A,B}, w_{C,A}, w_{B,C}, w_{A,B,C}\}$ denote their corresponding supports.*

This complex is also a pure 2-simplex, i.e. triangle, with one maximal independent connected component. The boundary of 2-H-simplex has four 0-faces (0-simplexes) and three 1-faces (1-simplexes). Since all the simplexes are in the complex, it is a closed complex. Therefore, we can say this complex represent a concrete concept. In general, the n-simplex has the following geometric property.

Property 1. The boundary of a n-H-simplex has $n+1$ 0-faces (vertices), $\frac{n(n+1)}{2}$ 1-faces (edges), and $\binom{n+1}{i+1}$ i-faces ($i \leq n$), where $\binom{n}{k}$ is a binomial coefficient.

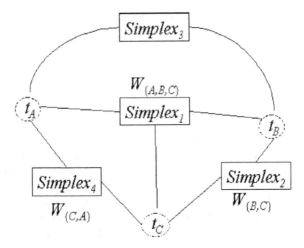

Fig. 2. A simple skeleton S_1^3 of example is composed of three terms $\{t_A, t_B, t_C\}$ from a collection of documents, where each simplex is identified by its tfidf value and all 0-simplices have been removed (the nodes are drawn by using dash circles)

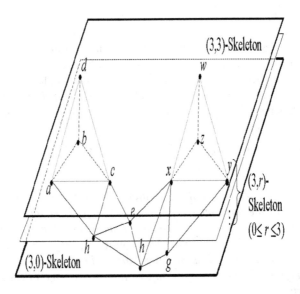

Fig. 3. A simple latent semantic network with its hierarchical structures is generated from Example 1. Obviously the skeleton $(3,3)$-Skeleton at the topmost layer composed of two maximal connected components as two distinct concepts Simplex(a, b, c, d) and Simplex(w, x, y, z) is contained in the skeleton at the lower layer. Except the topmost layer, all the concepts are in some sort of vague discrimination. The bottom layer contains only one connected component, which is a 3-complex. All the concepts are mixed together that make several primitive concepts are non-distinguishable in this connected component.

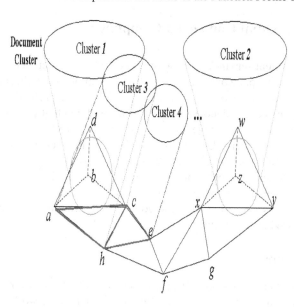

Fig. 4. Each cluster of documents is identified by a maximal connected component. Some cluster may overlap with other cluster because of the common face between them.

This geometric representation properly satisfies the *apriori* property of association rules: if the support of an item set $\{t_1, t_2, \cdots, t_n\}$ is bigger that a minimum support, so are all the non-empty subsets of it. In a complex, the universe of vertices organizes 1-simplices, i.e., frequent 1-itemsets, the universe of 1-simplex represents all possible frequent 1-itemsets and frequent 2-itemsets, and so on.

According to Example 1, it is obvious that simplices within the higher level skeleton S_r^n is contained in the lower level skeleton S_k^n with the same n-complex, $r \geq k$. Figure 3 shows the network hierarchy of the example, each skeleton is represented as a layer. For the purpose of simplicity, skeletons induced from r-complex, in which $0 \leq r < 3$, are neglected. The most distinct concepts of all (without a common concept between them) are existed in the topmost layer, although they could be empty concepts, which means there does not exist any non-overlapped concepts. In this example, the H-simplices $\mathrm{Simplex}(a, b, c, d)$ and $\mathrm{Simplex}(w, x, y, z)$ are two *maximal independent connected components* that demonstrate two discriminating primitive concepts. The H-simplices at the lower layers could have a common face between them. Therefore, the concepts denoted by those H-simplices are vague discriminated as shown in Figure 4 in that an overlapped concept induced by a common face is existed. As seen in the skeleton S_1^3, the maximal connected components generated from simplex $\mathrm{Simplex}(a, b, c, d)$ and simplex $\mathrm{Simplex}(a, c, h)$ have a common face $\mathrm{Simplex}(a, c)$ that makes some documents not able to properly discriminated in accordance with the generated association rules from term a and term c, so are the other maximal connected components in the skeleton. Because of the intersection produced by such subsimplices, some documents would be vague classified into two clusters. The lower the skeleton layer is, the serious the concept overlapping situation is.

4   Finding Maximal Connected Components

For the context of latent semantic ideas within a collection of documents, it is naturally that some similar concepts would be cross-referenced among the collection, especially for a collection of homogeneous documents. Therefore, some professional used words or phrases are often taken to denote a specific idea. No doubt that we can identify them by the usage of those terms. As we already known the best way to recognize them is according to term-term inter-relationships, which are term associations. Following the above statement, combinatorial geometry based latent semantic networks are the perfect model for illustrating the concepts in a huge variety of high-dimensional data, such a document collection. The algorithm for finding all concepts, i.e., maximal connected components, which is generated from the co-occurred terms in a collection of documents, will be introduced as follows.

4.1   Data Structure

In order for the further discussion on the algorithm, let us make the following definitions of the use of geometric notations to represent latent semantic networks on association rules.

Definition 9. *In a latent semantic network, let V be the set of single terms in a collection of documents, i.e., 0-simplices, and $\mathcal{E}$ be the set of all r-simplices, where $r \geq 0$. If $\mathrm{Simplex_A}$ is in $\mathcal{E}$, its support is defined as $w(\mathrm{Simplex_A})$, i.e., the tfidf of all terms in $\mathrm{Simplex_A}$ co-occurred in a collection of documents.*

A network, which is a complex in geometry, can be represented as a matrix.

Example 3. *As seen in Example 2, the 2-simplex of the network is the set $\{t_A, t_B, t_C\}$, which is also the maximal connected component that represents a primitive concept in a document collection. As Venn diagram, the incident matrix I and the weighted incident matrix I_W of the network are as follows.*

$$I = \begin{pmatrix} 1\,0\,1\,1 \\ 1\,1\,1\,0 \\ 1\,1\,0\,1 \end{pmatrix}.$$

$$I_W = \begin{pmatrix} w_{A,B,C} & 0 & w_{A,B} & w_{C,A} \\ w_{A,B,C} & w_{B,C} & w_{A,B} & 0 \\ w_{A,B,C} & w_{B,C} & 0 & w_{C,A} \end{pmatrix}.$$

The rows correspond to the terms and the columns correspond to the simplices.

Each simplex denotes a connected component, i.e., an undirected association rules. If the simplex is a maximal connected component, it defines a maximal frequent itemset. The number of terms in this connected component defines its *rank*, that is, if its rank is r it is equivalent to frequent $r + 1$-itemsets.

4.2   Algorithm

As we already known, a r-H-simplex denotes a r-connected component, which is a frequent $r + 1$-itemset. If we say a frequent itemset I_i identified by an H-simplex Simplex$_i$ is a subset of a frequent itemset I_j identified by Simplex$_j$, it means that Simplex$_i \subset$ Simplex$_j$. An H-simplex Simplex$_i$ is said to be a maximal connected component if no other H-simplex Simplex$_j \in \mathcal{E}$ is the superset of Simplex$_i$ for $i \neq j$. Documents can be automatically clustered based on all maximal connected components. It provide a soft-computing that allows overlapped concepts exist within a collection of documents.

All connected components are convex hulls, the intersection of connected components is nothing or a connected component. It would induce an vague region for concept discrimination if the intersection is a non-empty simplex. This common face will induce an unspecified concept be-tween them as we have mentioned before. It is not necessary to consider this common face because it has been considered in its super-simplices.

Example 4. *As shown in Figure 5, one component is organized by the H-simplex* Simplex$_1$={t$_A$, t$_B$, t$_C$}, *the other is generated by the H-simplex* Simplex$_5$={t$_C$, t$_D$, t$_E$}.
  The boundary of a concept defines all possible term associations in a document collection. Both of them share a common concept that can be taken as a 0-simplex {t$_C$}, *which is an 1-item frequent itemset* {t$_C$}.

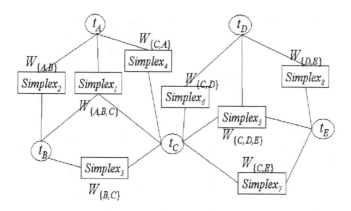

Fig. 5. A complex is composed of two maximal connected components generated by two 2-H-simplices Simplex(t$_A$, t$_B$, t$_C$) and Simplex(t$_C$, t$_D$, t$_E$). Both of them contain a common face Simplex(t$_C$) that produces an undiscriminating concept region.

Property 2. The intersection of concepts is nothing or a concept that is a maximal H-simplex belonging to all intersected concepts.

Since there is at most one maximal H-simplex in the intersection of more than one connected components and the dimension or rank of the intersection is lower than all intersected simplices. It is convenient for us to design an efficient algorithm for documents clustering based on all maximal connected components in a complex skeleton by skeleton. It does not need to traverse all complex.

5   Demonstration 1 - Graphical Display of Functions of the NOD2 Genes

Demonstration were performed on the relevant *GeneRIF (References into Function)* document set, publicly available in the EUtils web service of the NCBI Entrez site. Our geometrical model is suitable for representation of biological information, where hierarchical concepts in different complexity could be explored interactively according to the context of application and the various needs of the researchers.

The biological background of the experiment is briefly described here, with the terms or the concepts quoted. "CARD15" gene was found equivalent with "NOD2" gene in recent years. This CARD15/NOD2 gene was discovered associated with inflammatory bowel diseases ("IBDs") in 2000, and vigorous correlation studies were performed to elucidate the position on the genome or several candidate "chromosomes". The pathogenesis was proposed later to be "barrier" break in the intestinal ("mucosa") defense

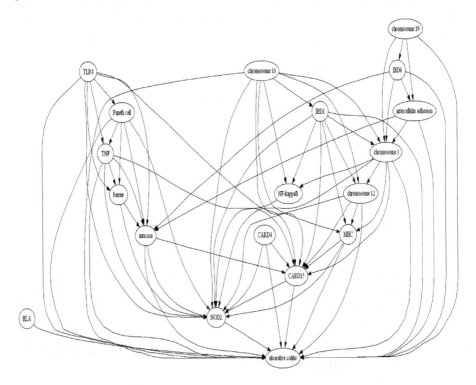

Fig. 6. Functional profiles of the CARD15 gene, rendered by Graph-Viz. The direction of edges are based on TFIDF weighting in this implementation. Our model does not imply directed association.

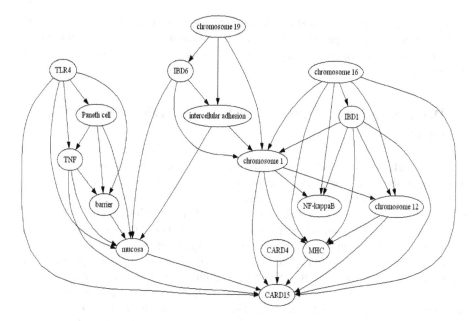

Fig. 7. Functional profiles of the CARD15 gene, with the threshold of the co-occurrence between concept raised. Three biologically meaningful clusters formed.

mechanism due to the genetic defect, then the focus of researchers shifted to the functioning domain of "inflammation" – "TNF", "TLR4", "NF-KappaB", and "Paneth cell".

The GIF document set of CARD15 gene was queried. The abstracts were retrieved, and the important keywords and synonyms were processed by a dictionary derived from UMLS thethaurus. The co-occurrences between the terms were calculated, weighted by TFIDF measurements. In this implementation, the term nodes were ranked by TFIDF weighting, and directed graphes were displayed for additional arrangement of the terms after suggestion by medical domain experts. Our model does not imply directed association.

The nodes of relevant concepts were rendered by the default setting of ATT GraphViz, the layout algorithm of which was according to geometrically even distribution of the nodes and their edges. The nodes with more interconnections or edges were positioneded together, compatible with the clusters of concepts in our model.

In Figure 6, the whole picture of term co-occurrence was shown. In Figure 7, the threshold of visible co-occurrence (the support) was raised, to show the 4-H-simplex or 5-H-simplex concept clusters. Three groups of 4-connected components or 5-connected components were shown in the left, the midlle, and the right regions, corresponding to the concept clusters of the new focus of "inflammatory process" and the older topics and genetic association and chromo-some localization.

The left "inflammatory process" cluster was the 5-frequent itemset with "TLR", "Paneth cell", "TNF", "barrier", and "mucosa". The middle and right clusters were two 4-H-simplex, connected by the intersection of the "chromosome 1" node.

A general concept is not good for classifying/clustering documents. A specific concept can achieve a good precision for document clustering, however, according to coordinate terms, some documents are unable to cluster into a same category by using different terms.

6   Demonstration 2 - Online "Clustering Search" of the NOD2 Gene

An online "clustering search" engine has been constructed based the methodology, which is available at http://ginni.bme.ntu.edu.tw. For use both on biomedical applications and on general purpose, our search engine receives query results (snippets) from PubMed or from Google. We described the PubMed "clustering search" in details, and those readers interested could verify our methodology by various general online searches. For feature extraction, a natural language processing package, QTAG, based on probabilistic parts-of speech natural language processing package was used, the phrase extraction were further fine-tuned by phrase patterns, stop words in some domains, and the weighing of the appearance in the title location. The extracted phrases were selected based on TFIDF measures. For example, the query of "NOD2" gene in PubMed returned 4831 phrases in 200 articles. After the selection by TF*IDF threshold of 0.1, 558 phrases remained to be clustered.

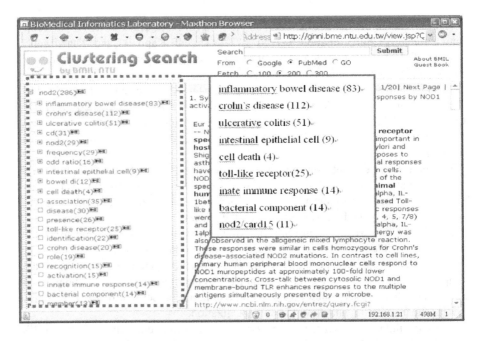

Fig. 8. The initial text browser that shows the hierarchical clustering results and the first nine clusters

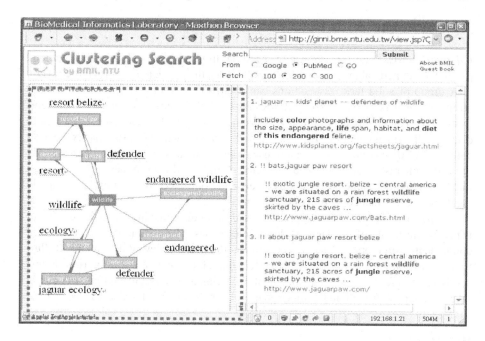

Fig. 9. Graph browser that shows the relations of features in the subgroup "leg ulcer"

Hierarchical clustering was achieved by association rules, which use support and confidence as similarity measure. Support denotes the ratio of the targets in the document corpus, and implies the significance of the association. Confidence is the support of antecedent and consequent divided by the support of the antecedent of the rule, and represents the comparative significance between two phrases. The assumption is that the documents of the same topic are expected to share more common itemsets, the simplicial complices in our model.

The algorithm is:

1. Define the initial value of support and confidence;
2. Find all association rules of pair features;
3. Choose the feature which is pointed by most other features, as the root of the subgroup;
4. Choose other features pointing to the root in this subgroup;
5. Recursively go to Step 2 until the number of the features in the subgroup is lower than a threshold.
6. Choose the feature pointed by most other features from the rest features in the pool, and go to Step 2 until there is no feature in the pool.

The result of the PubMed literature of the NOD2 gene was presented. Figure 8 was the initial screen showing the hierarchical results of clustering in text tree form.

In Figure 9, the graphical view of the details of network relationships between significant concepts in each cluster, user could drag the concept nodes of interest to change the focus and the arrangement to drill on the knowledge structure.

7 Conclusion

Polysemy, phrases and *term dependency* are the limitations of web search technology [5]. In the biomedical queries and concept analysis, the problem becomes more severe.

A group of solid term associations can clearly identify a concept. Most methods no matter what is *k-means, HCA, AutoClass* or *PDDP* classify or cluster documents from the represented matrix of a set of documents. It is inefficient and complicated to discover all term associations from such a high-dimensional and sparse matrix. Given a collection of documents, the associations among frequently co-occurring terms in any of the documents define naturally a geometric complex, which can then be decomposed into connected components at various levels and connected components can properly identify concepts in a collection of documents.

The paper presents a noval approach based on finding maximal connected components for clustering of the functional profile of genes. The r-simplexs, i.e., connected components, can represent the concepts in a collection of relevant documents. It illustrates that geometric complexes are a perfect model to denote association rules in text and is very useful for automatic document clustering and concept grouping, as demonstrated in our experiment in the functional analysis of gene-related documents.

References

1. GO Consortium. Go tools: Editors, browsers, general go tools and other tools. http://www.geneontology.org/doc/GO.tools.html, 2004.
2. C. Friedman, P. Kra, H Yu, M. Krauthammer, and A. Rzhetsky. Genies: a natural-language processing system for the extraction of molecular pathways from journal articles. *Bioinformatics*, 17(suppl 1):S74–82, 2001.
3. M. Girvan and M. Newman. Community structure in social and biological networks. In *Proceedings of the National Academy of Sciences*, volume 99, page 8271V76, 2002.
4. T. K. Jenssen, A. Laegreid, J. Komorowski, and E. Hovig. A literature network of human genes for high-throughput analysis of gene expression. *Nature Genetics*, 28(1):21V28, 2001.
5. A. Joshi and Z. Jiang. Retriever: Improving web search engine results using clustering. In A. Gangopadhyay, editor, *Managing Business with Electronic Commerce: Issues and Trends*, chapter 4. World Scientific, 2001.
6. P. Kankar, S. Adak, A. Sarkar, K. Murali, and G Sharma. Medmesh summarizer: Text mining for gene clusters. In *Proceedings of the Second SIAM International Conference on Data Mining*, Apr 2002.
7. J. R. Munkres. *Elements Of Algebraic Topology*. Addison Wesley, Reading MA, 1984.
8. H. Shatkay, S. Edwards, W. J. Wilbur, and M. Boguski. Genes, themes and microarrays: Using information retrieval for large-scale gene analysis. In *Proc Int Conf Intell Syst Mol Biol*, volume 8, pages 317–28, 2000.
9. J. D. Wren and H. R. Garner. Shared relationship analysis: ranking set cohesion and commonalities within a literature-derived relationship network. *Bioinformatics*, 20(2):191–8, 2004.

Evolutionary Construction of Granular Kernel Trees for Cyclooxygenase-2 Inhibitor Activity Comparison

Bo Jin[1] and Yan-Qing Zhang[2]

[1] Department of Computer Science, Georgia State University, Atlanta,
GA 30302-3994, USA
bojin@gsu.edu
[2] Department of Computer Science, Georgia State University, Atlanta,
GA 30302-3994, USA
yzhang@cs.gsu.edu

Abstract. With the growing interest of biological data prediction and chemical data prediction, more and more complicated kernels are designed to integrate data structures and relationships. We proposed a kind of evolutionary granular kernel trees (EGKTs) for drug activity comparisons [1]. In EGKTs, feature granules and tree structures are predefined based on the possible substituent locations. In this paper, we present a new system to evolve the structures of granular kernel trees (GKTs) in the case that we lack knowledge to predefine kernel trees. The new granular kernel tree structure evolving system is used for cyclooxygenase-2 inhibitor activity comparison. Experimental results show that the new system can achieve better performance than SVMs with traditional RBF kernels in terms of prediction accuracy.

1 Introduction

With the growing interests of biological data prediction and chemical data prediction such as structure-property based molecule comparison, protein structure prediction and long DNA sequence comparison, more complicated kernels are designed to integrate data structures, such as string kernels [2][3], tree kernels [4][5] and graph kernels [6][7]. The detailed review is given in [8]. One common character of these kernels is that feature transformations are implemented according to objects structures without steps of input feature generation. These transformations are very efficient in the case that objects include large structured information. While in many challenging problems, objects are not structured or some relationships within objects are not easy to be described directly.

We proposed a kind of evolutionary granular kernel trees (EGKTs) for drug activity comparisons [1]. In EGKTs, features within an input vector are grouped into feature granules according to the possible substituent locations of compounds. The similarity between two feature granules is measured by using a granular kernel, and all granular kernels are fused together by trees. The parameters of granular kernels and the connection weights of granular kernel trees (GKTs) are optimized by genetic algorithms (GAs). While sometimes due to the lack of prior knowledge, it would be

C. Priami et al. (Eds.): Trans. on Comput. Syst. Biol. V, LNBI 4070, pp. 25–35, 2006.
© Springer-Verlag Berlin Heidelberg 2006

hard to predefine kernel tree structures. Considering such kind of challenging problems, we in this paper present a new system to evolve the structures of GKTs. The new system is called granular kernel tree structure evolving system (GKTSES). In GKTSES, a population of granular kernel trees is first generated randomly. Crossover and mutation are then used to generate new populations of kernel trees. Finally a kernel tree with the best structure is selected for data classification. In applications, GKTSES is used for cyclooxygenase-2 inhibitor activity comparison. Experimental results show that the GKTSES can achieve better performance than GAs based SVMs with traditional RBF kernels in terms of prediction accuracy.

A genetic programming based kernel tree is presented in [9] in which input vectors as units can operate under sum, minus, magnitude or product operation. However, the approach does not guarantee that the obtained function is a kernel function.

The rest of the paper is organized as follows. The kernel definition and properties are given in Section 2. Section 3 describes the granular kernel concepts and properties. GKTSES is presented in Section 4. Section 5 shows the simulations of cyclo-oxygenase-2 inhibitor activity comparison. Finally, Section 6 gives conclusion and directs the future work.

2   Kernel Definition and Properties

Definition [2]: A kernel is a function K that for all $\vec{x}, \vec{z} \in X$ satisfies

$$K(\vec{x}, \vec{z}) = \langle \phi(\vec{x}), \phi(\vec{z}) \rangle \tag{1}$$

where ϕ is a mapping from input space $X = R^n$ to an inner product feature space $F = R^N$

$$\phi : \vec{x} \mapsto \phi(\vec{x}) \in F \tag{2}$$

The following are some popular kernel functions.

$$\text{Polynomial function } K(\vec{x}, \vec{y}) = (\vec{x} \bullet \vec{y} + 1)^d , \ d \in N \tag{3}$$

$$\text{RBF } K(\vec{x}, \vec{y}) = \exp(-\gamma \| \vec{x} - \vec{y} \|^2), \ \gamma \in R^+ \tag{4}$$

$$\text{Sigmoid kernel } K(\vec{x}, \vec{y}) = \tanh(\vec{x} \bullet \vec{y} - \theta) \tag{5}$$

Kernel properties [2]: If K_1 and K_2 are kernels on $X \times X$, the following $K(\vec{x}, \vec{y})$ are also kernel functions.

$$K(\vec{x}, \vec{y}) = cK_1(\vec{x}, \vec{y}), \ c \in R^+ \tag{6}$$

$$K(\vec{x}, \vec{y}) = K_1(\vec{x}, \vec{y}) + c , \ c \in R^+ \tag{7}$$

$$K(\vec{x}, \vec{y}) = K_1(\vec{x}, \vec{y}) + K_2(\vec{x}, \vec{y}) \tag{8}$$

$$K(\vec{x}, \vec{y}) = K_1(\vec{x}, \vec{y}) K_2(\vec{x}, \vec{y}) \tag{9}$$

3 Granular Kernel Concepts and Properties

Definition 1. A feature granule space G of input space $X = R^n$ is a sub space of X, where $G = R^m$ and $1 \le m \le n$.

Definition 2. A feature granule $\vec{g}$ is a vector which is defined in a feature granule space G.

Definition 3. A granular kernel gK is a kernel that for feature granules $\vec{g}, \vec{g}' \in G$ satisfies

$$gK(\vec{g}, \vec{g}') = \langle \varphi(\vec{g}), \varphi(\vec{g}') \rangle \tag{10}$$

where φ is a mapping from the feature granule space to an inner product feature space R^E.

$$\varphi : \vec{g} \mapsto \varphi(\vec{g}) \in R^E \tag{11}$$

Property 1. Granular kernels inherit the properties of traditional kernels such as the closure under sum, product, and multiplication with a positive constant over the granular feature spaces.

Property 2 [10][11]**.** A kernel can be constructed with two granular kernels defined over different granular feature spaces under sum operation.

To prove it, let $gK_1(\vec{g}_1, \vec{g}_1')$ and $gK_2(\vec{g}_2, \vec{g}_2')$ be two granular kernels, where $\vec{g}_1, \vec{g}_1' \in G_1$, $\vec{g}_2, \vec{g}_2' \in G_2$ and $G_1 \ne G_2$. We may define new kernels like this,
$gK((\vec{g}_1, \vec{g}_2), (\vec{g}_1', \vec{g}_2')) = gK_1(\vec{g}_1, \vec{g}_1')$
$gK'((\vec{g}_1, \vec{g}_2), (\vec{g}_1', \vec{g}_2')) = gK_2(\vec{g}_2, \vec{g}_2')$
gK and gK' can operate over the same feature space $(G_1 \times G_2) \times (G_1 \times G_2)$. We get
$gK_1(\vec{g}_1, \vec{g}_1') + gK_2(\vec{g}_2, \vec{g}_2') = gK((\vec{g}_1, \vec{g}_2), (\vec{g}_1', \vec{g}_2')) + gK'((\vec{g}_1, \vec{g}_2), (\vec{g}_1', \vec{g}_2'))$
According to the sum closure property (Eq. (8)) of kernels, $gK_1(\vec{g}_1, \vec{g}_1') + gK_2(\vec{g}_2, \vec{g}_2')$ is a kernel over $(G_1 \times G_2) \times (G_1 \times G_2)$.

Property 3 [10][11]**.** A kernel can be constructed with two granular kernels defined over different granular feature spaces under product operation.

To prove it, let $gK_1(\vec{g}_1, \vec{g}_1')$ and $gK_2(\vec{g}_2, \vec{g}_2')$ be two granular kernels, where $\vec{g}_1, \vec{g}_1' \in G_1$, $\vec{g}_2, \vec{g}_2' \in G_2$ and $G_1 \ne G_2$. We may define new kernels like this,

$$gK((\vec{g}_1, \vec{g}_2), (\vec{g}_1', \vec{g}_2')) = gK_1(\vec{g}_1, \vec{g}_1')$$
$$gK'((\vec{g}_1, \vec{g}_2), (\vec{g}_1', \vec{g}_2')) = gK_2(\vec{g}_2, \vec{g}_2')$$

So gK and gK' can operate over the same feature space $(G_1 \times G_2) \times (G_1 \times G_2)$. We get $gK_1(\vec{g}_1, \vec{g}_1')gK_2(\vec{g}_2, \vec{g}_2') = gK((\vec{g}_1, \vec{g}_2), (\vec{g}_1', \vec{g}_2'))gK'((\vec{g}_1, \vec{g}_2), (\vec{g}_1', \vec{g}_2'))$

According to the product closure property (Eq. (9)) of kernels, $gK_1(\vec{g}_1, \vec{g}_1')gK_2(\vec{g}_2, \vec{g}_2')$ is a kernel over $(G_1 \times G_2) \times (G_1 \times G_2)$.

According to the granular kernel properties, we can define kernels with various tree structures.

4   GKTSES

In GKTSES, a population of individuals is generated in the first generation. Each individual encodes a granular kernel tree. For example, three-layer GKTs (GKTs-1 and GKTs-2) are shown in Figure 1. In GKTs-1, each node in the first layer is a granular kernel. Granular kernels are combined together by sum and product connection operations in layer 2 and layer 3. Each granular kernel tree is encoded into a chromosome. For example, GKTs-1 and GKTs-2 are encoded in chromosomes c_1 and c_2 (see Figure 2) respectively. To generate an individual, features are randomly shuffled and then feature granules are randomly generated. Granular kernels are preselected from the candidate kernel set. Some traditional kernels such as RBF kernels and polynomial kernels can be chosen as granular kernels, since these kernels have proved successful in many real problems. In this paper, we choose RBF kernels as granular kernels and each feature as a feature granule. Finally granular kernel parameters and kernel connection operations are randomly generated for each individual.

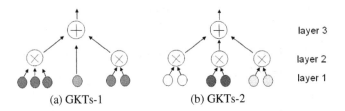

(a) GKTs-1   (b) GKTs-2

Fig. 1. Three-layer GKTs

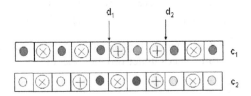

Fig. 2. Chromosomes

Chromosome: Let P_i denote the population in generation G_i, where $i = 1, \cdots m$ and m is the total number of generations. Each population P_i has p chromosomes $c_{ij}, j = 1, \cdots, p$. Each chromosome c_{ij} has $2q+1$ genes $g_t(c_{ij})$, where $t = 1, \cdots, 2q+1$. In each chromosome, genes $g_{2x-1}(c_{ij})$ $x = 1, \cdots, q+1$ represent granular kernels and genes $g_{2x}(c_{ij})$ $x = 1, \cdots, q$ represent sum or product operations. Here each granular kernel gene includes a random connection weight and kernel parameters, which are not evolved in the later optimization stages. We use $GKTs(c_{ij})$ to represent GKTs configured with genes $g_t(c_{ij}), t = 1, \cdots, 2q+1$.

Fitness: There are several methods to evaluate SVMs performance. One is using k-fold cross-validation, which is a popular technique for performance evaluation. Others are some theoretical bound evaluations on the generalization errors, such as Xi-Alpha bound [12], VC bound [13], Radius margin bound and VCs span bound [14]. Detail review can be found in [15]. In this paper we use k-fold cross-validation to evaluate SVMs performance in training phase.

In k-fold cross-validation, the training data set $\tilde{S}$ is separated into k mutually exclusive subsets $\tilde{S}_v$. For $v = 1, \cdots, k$, data set Λ_v is used to train SVMs with $GKTs(c_{ij})$ and $\tilde{S}_v$ is used to evaluate SVMs model.

$$\Lambda_v = \tilde{S} - \tilde{S}_v \ , v = 1, \cdots, k \tag{12}$$

After k times of training-testing on all different subsets, we get k prediction accuracies. The fitness f_{ij} of chromosome c_{ij} is calculated by

$$f_{ij} = \frac{1}{k} \sum_{v=1}^{k} Acc_v \tag{13}$$

where Acc_v is the prediction accuracy of $GKTs(c_{ij})$ on $\tilde{S}_v$.

Selection: In the algorithm, the roulette wheel method described in [16] is used to select individuals for the new population. Before selection, the best chromosome in generation G_{i-1} will replace the worst chromosome in generation G_i if the best chromosome in G_i is worse than the best chromosome in G_{i-1}. The sum of fitness values F_i in population G_i is first calculated.

$$F_i = \sum_{j=1}^{p} f_{ij} \tag{14}$$

A cumulative fitness $\tilde{q}_{ij}$ is then calculated for each chromosome.

$$\tilde{q}_{ij} = \sum_{t=1}^{j} \frac{f_{it}}{F_i} \qquad (15)$$

The chromosomes are then selected as follows. A random number r is generated within the range of [0, 1]. If r is smaller than $\tilde{q}_{i1}$, then chromosome c_{i1} is selected; otherwise chromosome c_{ij} is selected according to the following inequation.

$$\tilde{q}_{i\,j-1} < r \le \tilde{q}_{i\,j} \qquad (16)$$

After running the above select procedure p times, the new population is generated.

Crossover: Two GKTs are first selected from current generation as parents and then the crossover point is randomly selected to separate GKTs. Subtrees of two GKTs are exchanged at crossover point to generate two new GKTs. For example, two new GKTs are shown in Figure 3, which are generated from GKTs-1 and GKTs-2 through crossover operation at point d_2 (see Figure 2). In this example, two new GKTs (GKTs-3 and GKTs-4) have the same structures as their parents. In Figure 4, new GKTs-5 and GKTs-6 are generated from GKTs-1 and GKTs-2 through crossover operation at point d_1. The structures of two new GKTs in Figure 4 are different from those of GKTs in Figure 1. The related chromosomes of new GKTs are shown in Figure 3 (c) and Figure 4 (c).

In each population, one chromosome has the same probability to be selected to do crossover with another chromosome.

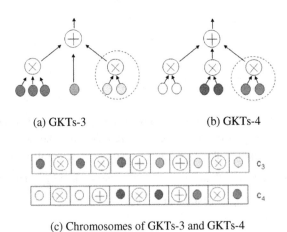

(a) GKTs-3   (b) GKTs-4

(c) Chromosomes of GKTs-3 and GKTs-4

Fig. 3. Crossover at point d_2

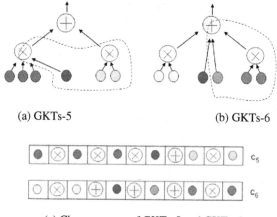

(a) GKTs-5   (b) GKTs-6

(c) Chromosomes of GKTs-5 and GKTs-6

Fig. 4. Crossover at point d_1

Mutation: In mutation, some genes of one chromosome are selected with the same probability. The values of selected genes are replaced by random values. In this paper, only connection operation genes are selected to do mutation. Figure 5 shows an example of mutation. In Figure 5 (a), the new chromosome c_7 is generated by changing the eighth gene of chromosome c1 from sum operation to product operation, which is equivalent to transforming GKTs-1 to GKTs-7 (see Figure 5 (b)).

The system architecture of GKTSES is shown in Figure 6. In the system, the regularization parameter C of SVMs is also optimized by GAs.

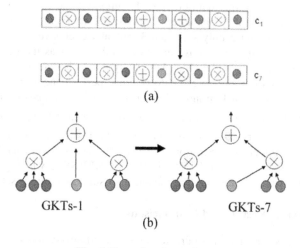

Fig. 5. Example of mutation

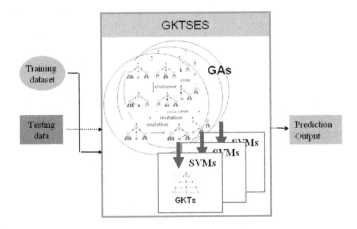

Fig. 6. System Architecture of GKTSES

5 Simulations

In simulations, GKTSES is compared to GAs based SVMs with traditional RBF kernels (which is called GAs-RBF-SVMs). In GAs-RBF-SVMs, the regularization parameter C and RBF's γ are optimized with GAs.

5.1 Dataset and Experimental Setup

The dataset of cyclooxygenase-2 inhibitors includes 314 compounds, which are described in [17]. The point of $\log(IC_{50})$ units used to discriminate active compounds from inactive compounds is set to 2.5. There are 153 active compounds and 161 inactive compounds in the dataset. 109 features are selected to describe each compound and each feature's absolute value is scaled to the range [0, 1]. The dataset is randomly shuffled and evenly split into 3 mutually exclusive parts. Each time one part is chosen as the unseen testing set and the other twos as the training set. Total three comparison results decide which kind of system is better. In the training phase, we use 3-fold cross-validation to evaluate systems' performances on training set. RBF kernel is also chosen as each granular kernel function. The ranges of all RBFs' γ are set to $[0.00001, 1]$ and the range of regularization parameter C is set to $[1, 256]$. The probability of crossover is 0.8 and the mutation ratio is 0.2. The population size is set to 300 and the number of generations is set to 50. The software package of SVMs used in the experiments is LibSVM [18]. In GKTSES, when generating each individual of the first generation, the probability of sum operation is set to 0.5.

5.2 Experimental Results and Comparisons

Table 1 shows the prediction accuracies of two kinds of systems in average. From Table 1, we can see that and GKTSES can outperform GAs-RBF-SVMs by 3.5% and

2% in terms of average testing accuracy and average fitness. The training accuracies of GAs-RBF-SVMs are higher than those of GKTSES. However, comparing training accuracy with the testing accuracy and the fitness in each system, we find overfitting happens in GAs-RBF-SVMs. So GKTSES is more reliable in the prediction.

Table 1. Prediction accuracies in average

	Fitness	Training accuracy	Testing accuracy
GAs-RBF-SVMs	81.9%	94.4%	72.3%
GKTSES	83.9%	89.7%	75.8%

Table 2. Prediction accuracies in three evaluations

	CV-1		CV-2		CV-3	
	GAs-RBF-SVMs	GKTSES	GAs-RBF-SVMs	GKTSES	GAs-RBF-SVMs	GKTSES
Fitness	83.7%	87.1%	80.4%	82.3%	81.4%	82.4%
Training accuracy	90.9%	92.3%	97.1%	88.0%	95.2%	88.6%
Testing accuracy	64.8%	68.6%	78.1%	81%	74%	77.9%
C	25.53	223.53	182.20	223.53	5.85	223.53
#Support vector	116	91	107	111	144	111

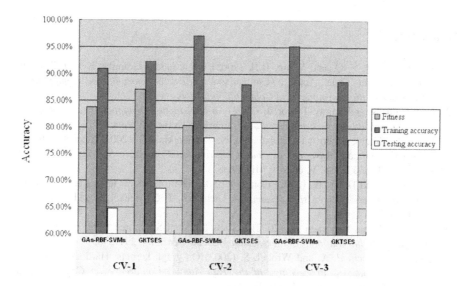

Fig. 7. Prediction accuracies in three evaluations

Table 2 shows the performances of two systems in detail. CV-1, CV-2 and CV-3 are three evaluations. From Table 2, we can see that the testing accuracies of GKTSES are always higher than those of GAs-RBF-SVMs by about 2.9% ~ 3.9% in three evaluations. The fitnesses of GKTSES are higher than those of GAs-RBF-SVMs by about 1% ~ 3.4%. We also find that the evolved values of regularization parameters (C) are same in three evaluations of GKTSES. So for the dataset of cyclooxygenase-2 inhibitors, 223.53 may be a good trade off point for GKTSES. Figure 7 shows the prediction accuracies of three evaluations in one picture.

6 Conclusion and Future Work

In this paper, we present a granular kernel tree structure evolving system to construct and evolve granular kernel trees. The new system is used for cyclooxygenase-2 inhibitor classification and the experimental results show that GKTSES can achieve better performances than GAs based SVMs with traditional RBF kernels. In the future, we will continue our research on the design of evolutionary granular kernel trees. Using genetic programming to construct granular kernel trees is one of research directions. We will also design parallel GAs to speed up evolving GKTs.

Acknowledgments

Special thanks to Dr. Peter C. Jurs and Dr. Rajarshi Guha for providing the dataset of cyclooxygenase-2 inhibitors. This work is supported in part by NIH under P20 GM065762. Bo Jin is supported by Molecular Basis for Disease (MBD) Doctoral Fellowship Program, Georgia State University.

References

1. Jin, B., Zhang, Y.-Q. and Wang, B.H. (2005) Evolutionary Granular Kernel Trees and Applications in Drug Activity Comparisons. *IEEE Symposium on Computational Intelligence in Bioinformatics and Computational Biology*, 121-126.
2. Cristianini, N. and Shawe-Taylor, J. (1999) *An introduction to support Vector Machines: and other kernel-based learning methods* Cambridge University Press, NY.
3. Lodhi, H., Shawe-Taylor , J., Christianini, N. and Watkins, C. (2001) Text classification using string kernels. *Advances in Neural Information Processing Systems* 13. Leen, T., Dietterich, T. and Tresp, V. editors, MIT Press.
4. Collins, M. and Duffy, N. (2002) Convolution kernels for natural language. *Advances in Neural Information Processing Systems* 14, Dietterich, T. G., Becker, S. and Ghahramani, Z. editors, MIT Press.
5. Kashima, H. and Koyanagi ,T. (2002) Kernels for Semi-Structured Data. *Proceedings of the Nineteenth International Conference on Machine Learning*, 291-298.
6. Gärtner, T. Flach, P. A. and Wrobel, S. (2003) On graph kernels: Hardness results and efficient alternatives. *Proceedings of the 16th Annual Conference on Computa tional Learning Theory and the 7th Kernel Workshop*.

7. Kashima, H. and Inokuchi, A. (2002) Kernels for graph classification. *In ICDM Workshop on Active Mining.*

8. Gärtner, Thomas (2003) A Survey of Kernels for Structured Data. *ACM SIGKDD Explorations Newsletter* 5, 49-58.

9. Howley, T. and Madden, M. G. (2004) The genetic evolution of kernels for support vector machine classifiers. *In Proceedings of 15th Irish Conference on Artificial Intelligence and Cognitive Science.*

10. Berg, C., Christensen, J. P. R. and Ressel, P. (1984) *Harmonic Analysis on Semigroups-Theory of Positive Definite and Related Functions*, Springer-Verlag.

11. Haussler, D. (1999) Convolution kernels on discrete structures. *Technical report UCSC-CRL-99-10, Department of Computer Science, University of California at Santa Cruz.*

12. Joachims, T. (2000) Estimating the Generalization Performance of a SVM Efficiently. *Proceedings of the International Conference on Machine Learning, Morgan Kaufman.*

13. Vapnik, V. N. (1998) *Statistical Learning Theory*, New York:John Wiley and Sons.

14. Vapnik, V.N. and Chapelle, O. (1999) Bounds on error expectation for support vector machine. *Advances in Large Margin ClassiBers.* Smola, A., Bartlett, P., SchRolkopf, B. and Schuurmans, D. editors , MIT Press, Cambridge, MA.

15. Duan, Kaibo, Keerthi, S. Sathiya and Poo, Aun Neow (2003) Evaluation of simple performance measures for tuning SVM hyperparameters. *Neurocomputing* 51, 41-59.

16. Michalewicz, Z. (1996) *Genetic Algorithms + Data Structures = Evolution Programs*, Springer Verlag, Berlin.

17. Kauffman, G. W. and Jurs, P. C. (2001) QSAR and k-Nearest Neighbor Classification Analysis of Selective Cyclooxygenase-2 Inhibitors Using Topologically-Based Numerical Descriptors. *J. Chem. Inf. Comput. Sci.* 41, 6:1553-1560.

18. Chang, C.-C. and Lin, C.-J. (2001) LIBSVM: a library for support vector machines. Software available at http://www.csie.ntu.edu.tw/~cjlin/libsvm.

A Control Model for Markovian Genetic Regulatory Networks

Michael K. Ng[1], Shu-Qin Zhang[2], Wai-Ki Ching[2], and Tatsuya Akutsu[3]

[1] Department of Mathematics, Hong Kong Baptist University,
Kowloon Tong, Hong Kong
`mng@math.hkbu.edu.hk`
[2] Department of Mathematics, The University of Hong Kong,
Pokfulam Road, Hong Kong
`sqzhang@hkusua.hku.hk, wkc@maths.hku.hk`
[3] Institute for Chemical Research, Kyoto University,
Gokasho Uji, Kyoto 611-0011, Japan
`takutsu@kuicr.kyoto-u.ac.jp`

Abstract. In this paper, we study a control model for gene intervention in a genetic regulatory network. At each time step, a finite number of controls are allowed to drive to some target states (i.e, some specific genes are on, and some specific genes are off) of a genetic network. We are interested in determining a minimum amount of control cost on a genetic network over a certain period of time such that the probabilities of obtaining such target states are as large as possible. This problem can be formulated as a stochastic dynamic programming model. However, when the number of genes is n, the number of possible states is exponentially increasing with n, and the computational cost of solving such stochastic dynamic programming model would be very huge. The main objective of this paper is to approximate the above control problem and formulate as a minimization problem with integer variables and continuous variables using dynamics of states probability distribution of genes. Our experimental results show that our proposed formulation is efficient and quite effective for solving control gene intervention in a genetic network.

1 Introduction

Probabilistic Boolean networks (PBNs) have been proposed to study the dynamic behavior of gene regulatory networks [8]. It is a generalization of the standard Boolean networks. A Boolean network $G(V, F)$ consists of a set of nodes:

$$V = \{v_1, v_2, \ldots, v_s\}$$

and $v_i(t)$ represents the state (0 or 1) of v_i at time t. A list of Boolean functions:

$$F = \{f_1, f_2, \ldots, f_s\}$$

representing rules of regulatory interactions among the nodes (genes):

$$v_i(t+1) = f_i(\widehat{v}(t)), \quad i = 1, 2, \ldots, s,$$

C. Priami et al. (Eds.): Trans. on Comput. Syst. Biol. V, LNBI 4070, pp. 36–48, 2006.

where $\widehat{v}(t) = [v_1(t), v_2(t), \ldots, v_s(t)]^T$. The Boolean network is a deterministic model. However, gene expression is stochastic in nature and there is also experimental noise due to complex measurement process. To overcome the deterministic rigidity of a Boolean network, extension to a probabilistic setting is necessary. In a PBN, for each node, instead of having one Boolean function, there are a number of predictor functions $f_j^{(i)}$ for determining the state of gene v_i if it is chosen. For each gene v_i, $l(i)$ is the number of possible predictor functions and $c_j^{(i)}$ is the probability that $f_j^{(i)}$ is being chosen, and it is estimated by using Coefficient of Determination (COD)[4]. By incorporating more possible Boolean functions into each gene, they are able to cope with the uncertainty, which is intrinsic to biological systems. At the same time, they also share the appealing rule-based properties of standard Boolean networks [5,6,7]. The dynamics of PBNs also can be understood from the Markov chain point of view. Thus the numerous theories for Markov chains can also be applied to analyze the PBNs.

Although a PBN allows for uncertainty of inter-gene relations during the dynamic process, it will evolve only according to certain fixed transition probabilities. There is no mechanism for controlling this evolution towards more desirable states. To facilitate PBNs to evolve towards some given desired directions, intervention has been studied in some different ways. It has been shown that given a target state, one can facilitate the transition to it by toggling the state of a particular gene from on to off or vice-versa [9]. However, making a perturbation or a forced intervention can only be applied at one time point. The behavior of the system thereafter still depends on the network itself. The network may eventually return to some undesirable state after many steps. Another way is by using structural intervention to change the stationary behavior of the PBNs [10]. This approach also constitutes transient intervention. Since it involves the structural intervention, it is more permanent than the first one.

To increase the likelihood of transitions from an undesirable state to a desirable one in a PBN, more auxiliary variables can be involved in the system. Such variables are called control inputs. They take the binary values: 0 or 1, which indicates that a particular intervention is ceased or actively applied. The control can be applied in finite steps, not only at one time point. In [3], the control problem is formulated as a minimization problem of some costs. Under the supervision of biologists or clinicians, the cost functions are defined as the cost of applying the control inputs in some particular states. For the terminal states, all possible states are assumed to be reachable. Higher terminal costs are assigned to the undesirable states. Then, the control problem is to minimize the total cost under the condition that each step evolution is based on the transition probability which now is a function with respect to the control inputs. Since the system is stochastic in nature, the cost is given by its expectation. The optimal control problem is solved by the technique of stochastic dynamic programming. The simulations for this model indicate that the final state will be the desirable state with higher probability when using controls. For more details, we refer readers to the paper by Datta et al. [3]. We remark that when the number of genes is n, the number of possible states is exponentially increasing with n

(for instance, it is equal to 2^n when either on or off is used to describe the status of a gene). It implies that the computational cost of solving this stochastic dynamic programming model would be very huge. In [1], we have shown that finding a control strategy for Boolean network to the desired global state is NP-hard. This result justifies existing exponential time algorithms for finding control strategies for probabilistic Boolean netowrks.

In this paper, we approximate and formulate the gene intervention problem with a control model which is easy to understand and implement. At each time step, a finite number of controls can be put to drive the states to the desirable ones. The objective is to achieve a target state probability distribution with a minimal control cost. The model is formulated as a minimization problem with integer variables and continuous variables. There are many methods to solve such problems [11]. We use LINGO, a popular software for solving such minimization problem, to get control solutions for gene intervention of a genetic network.

The remainder of the paper is organized as follows. In Section 2, we give a brief review on PBNs and in Section 3, we formulate the linear control problem. In Section 4, experimental results are given to demonstrate the effectiveness of the linear control models and the efficiency of our algorithms. Finally, concluding remarks are given to discuss further research issues in Section 5.

2   The Mathematical Formulation

2.1   Review of the Probabilistic Boolean Network

In this section, we first review the PBNs briefly, we then present a previous control model. We are interested in modeling the relationship among "n" genes. In such a genetic network, each gene can take one of the two binary values: 0 or 1, or one of the three ternary values: -1, 0 or 1. For the former case, 0 and 1 correspond to the case that a particular gene is not expressed and expressed. For the latter case, -1, 0 and 1 indicate that the gene is down-regulated, unchanged and up-regulated respectively. Here we assume that each gene takes binary values in the discussion.

Suppose that the activity level of gene "i" at time step "k" is denoted by $x_i(k)$ where $x_i(k) = 0$ or 1. The overall expression levels of all the genes in the network at time step k is given by the following column vector

$$\widehat{x}(k) = [x_1(k), x_2(k), \ldots, x_n(k)]^T.$$

This vector is referred to the *Gene Activity Profile* (GAP) of the network at time k. For $\widehat{x}(k)$ ranging from $[0, 0, \ldots, 0]^T$ (all entries are 0) to $[1, 1, \ldots, 1]^T$ (all entries are 1), it takes on all the 2^n possible states of the n genes.

Furthermore, for the ith gene, there corresponds $l(i)$ possible Boolean functions:

$$\left\{ f_j^{(i)} : \text{ for } j = 1, \ldots, l(i) \right\},$$

and the probability of selecting function $f_j^{(i)}$ is $c_j^{(i)}$, where $f_j^{(i)}$ is a function with respect to the activity levels of n genes. Since $c_j^{(i)}$ are probabilities, they must satisfy the following condition:

$$\sum_{j=1}^{l(i)} c_j^{(i)} = 1.$$

For such a PBN with n genes, there are at most

$$N = \prod_{i=1}^{n} l(i)$$

different Boolean networks among these n genes. This means that there are totally N possible realizations of genetic networks. Let f_j be the jth possible realization,

$$f_j = [f_{j_1}^{(1)}, f_{j_2}^{(2)}, \ldots, f_{j_n}^{(n)}], \quad 1 \le j_i \le l(i), \quad i = 1, 2, \ldots, n.$$

Suppose that P_j is the probability of choosing the jth Boolean network,

$$P_j = \prod_{i=1}^{n} c_{j_i}^{(i)}, \quad 1, 2, \ldots, N. \tag{1}$$

Let $\widehat{a}$ and $\widehat{b}$ be any two column vectors with n entries being either 0 or 1. Then

Prob $\{\widehat{x}(k+1) = \widehat{a} \mid \widehat{x}(k) = \widehat{b}\}$

$$= \sum_{i=1}^{N} \text{Prob } \{\widehat{x}(k+1) = \widehat{a} \mid \widehat{x}(k) = \widehat{b}, \text{ where } i\text{th network is selected } \} \cdot P_i.$$

$$\tag{2}$$

By letting $\widehat{a}$ and $\widehat{b}$ ranging from $[0, 0, \ldots, 0]^T$ to $[1, 1, \ldots, 1]^T$ independently, we can get the transition probability matrix A. For the ease of presentation, we first transform the n-digit binary number vector, as discussed in [8], into a decimal number by the following formulae:

$$y(k) = 1 + \sum_{j=1}^{n} 2^{n-j} x_j(k).$$

As $\widehat{x}(k)$ ranges from $[0, 0, \ldots, 0]^T$ to $[1, 1, \ldots, 1]^T$, $y(k)$ will cover all the values from 1 to 2^n. Since the mapping from $\widehat{x}(k)$ to $y(k)$ is one-to-one, we can just equivalently work with $y(k)$.

Let $\mathbf{w}(k)$ be the probability distribution vector at time k, i.e.,

$$[\mathbf{w}(k)]_i \equiv w_i(k) = \text{Prob } \{y(k) = i\}, \quad i = 1, 2, \ldots, 2^n.$$

It is straightforward to check that

$$\mathbf{w}(k+1) = A\mathbf{w}(k). \tag{3}$$

where A satisfies

$$\sum_{i=1}^{2^n}[A]_{ij} = 1$$

and it has at most $N \cdot 2^n$ non-zero entries of the 2^n-by-2^n transition matrix.

2.2   A Previous Control Model

Suppose that there are m possible control inputs that can be applied to the PBNs at each time step. At each time step k,

$$\widehat{u}(k) = [u_1(k), u_2(k), \ldots, u_m(k)]^T$$

indicates the control status. As in the PBNs, $\widehat{u}(k)$ can take all the possible values from $[0, 0, \ldots, 0]^T$ to $[1, 1, \ldots, 1]^T$. One can still represent the controls with the decimal numbers

$$s(k) = 1 + \sum_{i=1}^{m} 2^{m-i} u_i(k).$$

As $\widehat{u}(k)$ ranges from $[0, 0, \ldots, 0]^T$ to $[1, 1, \ldots, 1]^T$, $s(k)$ can cover all the values from 1 to 2^m.

In [3], after applying the controls to the PBNs, the one-step time evolution of the probability distribution vector follows the equation:

$$\mathbf{w}(k+1) = A(s(k))\mathbf{w}(k). \tag{4}$$

We see that the probability distribution vector does not depend only on the previous probability distribution but also on the controls at each time step. By appropriately choosing the control inputs, the states of the network can be led to a more desirable status. The control problem is then formulated as follows. Given an initial state $y(0)$, find a control law

$$\pi = \{\widehat{u}(0), \widehat{u}(1), \ldots, \widehat{u}(M-1)\}$$

or

$$\pi = \{s(0), s(1), \ldots, s(M-1)\}$$

that minimizes the cost function:

$$J(\pi) = \mathcal{E}\left[\sum_{k=0}^{M-1} C_k(y(k), \widehat{u}(k)) + C_M(y(M))\right] \tag{5}$$

subject to

$$\text{Prob}\,\{y(k+1) = j \mid y(k) = i\} = a_{ji}(s(k)). \tag{6}$$

Here $C_k(y(k), \widehat{u}(k))$ (or $C_k(y(k), s(k))$ are the costs of applying the control input $\widehat{u}(k)$ $(s(k))$ when the state is $y(k)$. The optimal solution of this problem is given by the last step of the following dynamic programming algorithm which proceeds backward in time from time step $M - 1$ to time step 0:

$$
\begin{cases}
J_M(y(M)) = C_M(y(M)) \\
J_k(y(k)) = \min\limits_{s(k) \in \{1,2,\ldots,2^m\}} \mathcal{E}[C_k(y(k), s(k)) + J_{k+1}(y(k+1))], \\
\qquad \text{or} \\
J_k(y(k)) = \min\limits_{\widehat{u}(k) \in \{[0,0,\cdots,0]^T, \cdots, [1,1,\cdots,1]^T\}} \mathcal{E}[C_k(y(k), \widehat{u}(k)) + J_{k+1}(y(k+1))], \\
\qquad\qquad\qquad\qquad\qquad\qquad k = 0, 1, 2, \ldots, M - 1.
\end{cases}
$$

For more details, we refer readers to Datta et al.[3].

The above control problem is to put the controls on the transition probability matrix in each time step, such that the system can evolve towards the more desirable states.

3   The Proposed Model

In [1], we have shown that finding a control strategy for Boolean network to the desired global state is NP-hard. This result justifies existing exponential time algorithms for finding control strategies for probabilistic Boolean netowrks. Thus, the dynamic programming algorithm can only be applied to small biological systems. The main aim of this paper is to approximate the above control model and formulate it as an effective minimization problem such that the control on a larger genetic regulatory network can be studied.

We consider a discrete linear control system:

$$
\mathbf{w}(k + 1) = \alpha_k A \mathbf{w}(k) + \beta_k B \widehat{u}(k). \tag{7}
$$

Here we use the same notations as in Section 2, α_k and β_k are two parameters (to be discussed later) and the matrix B is the control transition matrix. It can be set in each column to represent the transition from one specific state to another on a particular gene. For example, we can set in the first column such that the first gene makes a transition from 0 to 1, then the first 2^{n-1} entries are 0 and the others are nonzero with the sum being equal to one in this column. Through the matrix B, the controls are effectively transferred to different possible states in the PBN. It is clear that $u_i(k) = 1$ means the active control is applied at the time step k while $u_i(k) = 0$ indicates that the control is ceased. If there are m possible controls at each time step, then the matrix B is of the size: $2^n \times m$.

Starting from the initial state or initial state probability distribution $\mathbf{w}(0)$, one may apply the controls

$$
\widehat{u}(0), \widehat{u}(1), \ldots, \widehat{u}(k - 1)
$$

to drive the probability distribution of the genetic network to some desirable state probability distribution at time step k. The evolution of the system now depends on both the initial state probability distribution and the controls in each time step. To construct $w(k)$ to be a probability distribution, we consider

$$\alpha_k + \beta_k = 1.$$

When there is no control at step k, we set

$$\alpha_k = 1 \quad \text{and} \quad \beta_k = 0.$$

Here β_k refers to the intervention ability of the control in a genetic regulatory network.

We remark that the traditional discrete linear control problem [2] does not have such parameters. The main reason is that in the traditional control problem, $\mathbf{w}(k)$ is the state (described by a numerical value, not a probability) of a system. However, $\mathbf{w}(k)$ is a probability distribution in this paper. We need to make sure that starting from the initial probability distribution, one applies controls to drive the probability distribution of the system to become some particular target probability distribution at the time instance k.

Given the objective state or state probability distribution at time k, we aim at finding the optimal controls:

$$\widehat{u}^*(0), \widehat{u}^*(1), \ldots, \widehat{u}^*(k-1),$$

such that the target probability distribution based on (5) can be maximized. In our setting, we apply at most one control at each time step. This means that the total number of nonzero $u_i(k)$ at each time step should be at most one.

According to the cost function in (5), we have

$$\mathcal{E}\left[\sum_{k=0}^{M-1} C_k(y(k), \widehat{u}(k)) + C_M(y(M))\right]$$

$$= \sum_{k=0}^{M-1} \mathcal{E}[C_k(y(k), \widehat{u}(k))] + \mathcal{E}[C_M(y(M))]$$

$$= \sum_{k=0}^{M-1} \sum_{i=1}^{2^n} \operatorname{Prob}(y(k) = i) C_k(y(k) = i, \widehat{u}(k)) + \sum_{i=1}^{2^n} \operatorname{Prob}(y(M) = i) C_M(y(M) = i)$$

$$= \sum_{k=0}^{M-1} \sum_{i=1}^{2^n} w_i(k) C_k(y(k) = i, \widehat{u}(k)) + \sum_{i=1}^{2^n} w_i(M) C_M(y(M) = i).$$

Therefore the following optimization problem is derived:

$$\min_{\widehat{u}(0), \widehat{u}(1), \cdots, \widehat{u}(M-1)} \left\{ \sum_{k=0}^{M-1} \sum_{i=1}^{2^n} w_i(k) C_k(y(k) = i, \widehat{u}(k)) + \sum_{i=1}^{2^n} w_i(M) C_M(y(M) = i) \right\}$$

subject to

$$\mathbf{w}(k+1) = \alpha_k A \mathbf{w}(k) + \beta_k B \widehat{u}(k), \quad k = 0, 1, \ldots, M-1,$$
$$w_i(k) = [\mathbf{w}(k)]_i \quad i = 1, 2, \ldots, 2^n, \ k = 0, 1, \ldots, M-1,$$
$$\alpha_k + \beta_k = 1, \quad k = 0, 1, \ldots, M-1,$$
$$\widehat{u}(k) = [u_1(k), u_2(k), \ldots, u_m(k)]^T, \quad k = 0, 1, \ldots, M-1,$$
$$u_i(k) \in \{0, 1\}, \quad i = 1, 2, \ldots, m, \ k = 0, 1, \ldots, M-1,$$
$$\sum_{i=1}^{m} u_i(k) \le 1, \quad k = 0, 1, \ldots, M-1,$$
$$\alpha_k = 1 - \beta_k \sum_{i=1}^{m} u_i(k), \quad k = 0, 1, \ldots, M-1.$$

The last constraint is to guarantee that if there is no control (i.e., $\sum_{i=1}^{m} u_i(k) = 0$), α_k should be equal to 1. Here β_k is a parameter for the intervention ability of the control in a genetic regulatory network.

The above formulation can be seen as only involving $m \times M$ integer variables $u_i(k)$ and linear constraints, which constitutes an Integer Programming (IP) model. Practically, both the number of time steps and the number of the controls are not very large (i.e., m and M are quite small), the total number of integer variables is not large. We expect that the computational times required for solving IP model is not huge. In the next section, we demonstrate this IP model is quite efficient to solve.

4   Experimental Results

In this section, we present two examples to show optimal design with integer programming model. The first one is to illustrate how our method can be applied and the second one is based on a more complex genetic network.

4.1   A Simple Example

In this subsection, we present an example to illustrate how an integer programming model can be applied to get the optimal control strategy. It is exactly the same example as that in [3]. The example involves a PBN with three genes: x_1, x_2 and x_3. There are two boolean functions $f_1^{(1)}$, $f_2^{(1)}$ associated with x_1, one boolean function $f_1^{(2)}$ associated with x_2, and two boolean functions $f_1^{(3)}$, $f_2^{(3)}$ associated with x_3. These boolean functions are given by the truth table in Table 1.

In this example, we try to control x_1 which is same as that in [3]. To make a clear comparison with the dynamic programming model, we make the same setting. The control variable is defined as gene x_1, which now is u. When there is no control, x_2 and x_3 will evolve according to the transition matrix A, which

Table 1. The truth table of boolean functions

$x_1\ x_2\ x_3$	$f_1^{(1)}$	$f_2^{(1)}$	$f_1^{(2)}$	$f_1^{(3)}$	$f_2^{(3)}$
0 0 0	0	0	0	0	0
0 0 1	1	1	1	0	0
0 1 0	1	1	1	0	0
0 1 1	1	0	0	1	0
1 0 0	0	0	1	0	0
1 0 1	1	1	1	1	0
1 1 0	1	1	0	1	0
1 1 1	1	1	1	1	1
$c_j^{(i)}$	0.6	0.4	1	0.5	0.5

can be got from the probabilistic boolean network when $x_1 = 0$ in Table 1. It is the transpose of the matrix $A(1)$ in [3]:

$$A = A(1)^T = \begin{pmatrix} 1 & 0 & 0 & 0.5 \\ 0 & 0 & 0 & 0.5 \\ 0 & 1 & 1 & 0 \\ 0 & 0 & 0 & 0 \end{pmatrix}, \qquad A(2)^T = \begin{pmatrix} 0 & 0 & 0.5 & 0 \\ 0 & 0 & 0.5 & 0 \\ 1 & 0.5 & 0 & 0 \\ 0 & 0.5 & 0 & 1 \end{pmatrix}$$

Next the task is to construct the control transition matrix B. The entries of B correspond to change of probabilities of all the states when the controls are applied. In this example, we only apply one control, so the matrix B has only one column. Since matrix $A(2)^T$ is the transition matrix when the control is applied ($u = 1$ for x_1), B can be constructed from $A(2)^T$. The change of the probability of state j when the control is applied is the contribution of transitions from all the states. It is the sum of all the entries that denote the transition to state j. To make the sum of all the entries in B be one, we need to make a normalization. Thus,

$$B(j) = \frac{1}{4} \sum_i [A(2)]_{i,j},$$

Here,

$$B(1) = \frac{0.5}{4} = 0.125, \quad B(2) = \frac{0.5}{4} = 0.125$$

$$B(3) = \frac{1 + 0.5}{4} = 0.375, \quad B(4) = \frac{1 + 0.5}{4} = 0.375$$

After the matrices A and B are determined, we need to formulate the problem. The control action is to be carried out over five steps [3]. The terminal penalties are given by $C(1) = 0, C(2) = 1, C(3) = 2, C(4) = 3$. The cost of the control at a certain state is 1.

By solving the integer programming problem, we can get the optimal control strategy for this problem.

- Case (1) The initial state is the first state $[0,0]$: According to the computation, the optimal control strategy in this case is no control, which is same as that in the paper [3]. The value of the optimal control cost is 0. The genes are always in the state [0,0].

- Case (2) The initial state is the fourth state [1,1]: In this case, the evolution of the PBN is starting from the most undesirable state. With the computed control strategy, we need not apply any control and the system will evolve to the first state with the probability 0.5.

To compare the proposed method with the dynamic programming method for this problem, we give the evolution of the system in Table 2 when there is no control applied. In Table 3, we present the control results when we take different values of β_k. We find that the larger the value of β_k is, the better results we have. We recall that β_k is a parameter for the intervention ability of the control in a genetic regulatory network. The large value of β_k means the high intervention

Table 2. The evolution of the system without controls

initial state	[0,0]	[0,1]	[1,0]	[1,1]
J_M	0	2	2	1
$w(5)$	[1, 0, 0, 0]	[0, 0, 1, 0]	[0, 0, 1, 0]	[0.5, 0, 0.5, 0]

Table 3. The control results using the proposed control model

	all β_k, if $\beta_k \neq 0$	initial state			
		[0,0]	[0,1]	[1,0]	[1,1]
$w(5)$	0.1	[1,0,0,0]	[0.06,0.04,0.90,0]	[0.06,0.04,0.90,0]	[0.50,0,0.50,0]
J_M	0.1	0	2.84	2.84	1
$w(5)$	0.2	[1,0,0,0]	[0.13,0.07,0.80,0]	[0.13,0.07,0.80,0]	[0.50,0,0.50,0]
J_M	0.2	0	2.67	2.67	1
$w(5)$	0.4	[1,0,0,0]	[0.16,0.09,0.75,0]	[0.16,0.09,0.75,0]	[0.50,0,0.50,0]
J_M	0.4	0	2.59	2.59	1
$w(5)$	0.6	[1,0,0,0]	[0.19,0.11,0.70,0]	[0.19,0.11,0.70,0]	[0.50,0,0.50,0]
J_M	0.6	0	2.51	2.51	1
$w(5)$	0.8	[1,0,0,0]	[0.25,0.15,0.60,0]	[0.25,0.15,0.60,0]	[0.50,0,0.50,0]
J_M	0.8	0	2.35	2.35	1
$w(5)$	0.99	[1,0,0,0]	[0.31,0.19,0.50,0]	[0.31,0.19,0.50,0]	[0.50,0,0.50,0]
J_M	0.99	0	2.19	2.19	1
u		no control	fourth step	fourth step	no control

Table 4. The control results using the dynamic programming model

initial state	[0,0]	[0,1]	[1,0]	[1,1]
J_M	0	1.5	1.5	1.25
$w(5)$	[1,0,0,0]	[0.50,0.50,0,0]	[0.50,0.50,0,0]	[0.75,0.25,0,0]
u	no control	fifth step	fifth step	fifth step

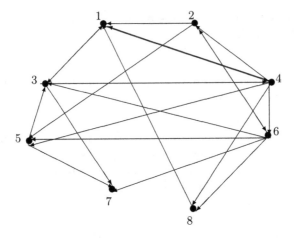

Fig. 1. The probabilistic boolean network of the eight genes

ability of the control in the network. The probability at the first state after we apply the controls, cannot be larger than a certain value. This originates from the matrix B, which is determined by the system itself. Table 4 is the results when the dynamic programming is applied.

4.2 A More Complex Example

In this subsection, we present an example to demonstrate the efficiency of the optimal control design by the integer linear programming approach. In this example, we consider a PBN of eight genes, x_1, x_2, ..., x_8. For each gene i, we assume that it can take two values: 0 or 1. Here 1 means the gene is expressed and 0 means it is not expressed.

We further assume that there are two probabilistic boolean functions: $f_1^{(i)}$ and $f_2^{(i)}$ associated with each gene i. All the probabilistic boolean functions and their variables are generated randomly. At the same time, the probability of the two boolean functions being applied to the corresponding particular gene is obtained. Figure 1 shows the network of these eight genes. Suppose that in this example, we expect the gene 1 is not expressed. Then controls will be introduced to drive the gene 1 from state 1 to state 0. Before solving the optimization problem formulated in the previous section, we need to do the following two steps:

- (a1) Obtain the matrices A and B, where A and B are the corresponding transition matrix and the control matrix respectively. Since there are two boolean functions for each gene, there are totally 2^8 possible networks to be considered. From (1), the probability of choosing any one of all the 2^8 can be obtained. By (2), we can get matrix A. To construct matrix B, in practice we need the opinions from the biologists to determine which gene can be easily controlled or have close relation with the target gene. For the purpose of demonstration, we will control gene 1 through all the eight genes and let them move from state 1 to state 0 with equal probability.

- (a2) Determine the cost of controls and the penalty for the states. We assign a cost of 1 to each forcible control. For the states penalty, since we expect the gene 1 to be in state 0, we assign a penalty of 0 to the states for which gene 1 equals 0 and a penalty of 3 to all the states for which gene 1 equals 1. We choose the penalty and cost arbitrarily. In practice, we still need some criteria to determine these costs and penalties.

Now we solve our optimization problem which is an integer programming problem. In this example, for all the β_k, if $\beta_k \neq 0$, $\beta_k = 0.5$. We choose the control such that it is only applied in three steps: $0, 1, 2$. With the popular software LINGO, we can get the solution in about one minute. The following are some results for initial state being both desirable and undesirable. It is clear to see the effect of this control strategy.

- (b1) The initial state is $[0, 0, 0, 0, 0, 0, 0, 0, 0]$, which is the desirable state. If we do not apply any control, the probability of this state evolving to the state for which gene 1 equals 0 after three steps is 0.2493. Under the control strategy, the probability is 0.6571, which is much higher than that without any control. The control strategy is as follows: in the first step we control gene 3; in the second step, we control gene 8 and in the third step, we control gene 1 again. As we have assumed, all the corresponding genes will be made to change from 1 to 0.
- (b2) The initial state is $[1, 1, 1, 0, 1, 0, 1, 1]$, which is not the desirable state. This time if we do not apply any control, the final state will be the desirable state with the probability 0.1913, which is a very small likelihood. However, with the optimal control strategy, the state will evolve to the desirable state with probability 0.6498. In this case, the control strategy is same as above.

In this example, we find that our approach takes less than 1 minute to compute the control solution under the PC with Platinum M and 256 kB RAM. And if the number of controls is not restricted to one, the computational time is still less than 1 minute.

5   Concluding Remarks

In this paper, we introduced a linear control model with the general control model form based on the PBN model of gene regulatory networks. At each step, one or more controls can be put to drive the genes to more desirable states. The control strategy can be used in the real life for therapeutic intervention. The optimal control results presented in this paper assume that the control transition matrix is known or just get it from the transition matrix when there is no any control. To get a reasonable control transition matrix is our further research topic.

References

1. Akutsu, T., Hayashida, M., Ching, W., Ng, M.: On the Complexity of Finding Control Strategies for Boolean Networks, The Fourth Asia Pacific Bioinformatics Conference 13-16 Feb, Taiwan (2006)
2. Chui, C., Chen, G. : Linear Systems and Optimal Control. Springer-Verlag, New York (1989)
3. Datta, A. Bittner, M.L., Dougherty, E.R.: External control in Markovian genetic regulatory networks. Machine Learning **52** (2003) 169-191
4. Dougherty, E.R. , Kim, S., Chen, Y.: Coefficient of determination in nonlinear signal processing. Signal Processing **80** (2000) 2219-2235
5. Kauffman, S.A.: Metabolic stability and epigenesis in randomly constructed genetic nets. Theoretical Biology **22** (1969) 437-467
6. Kauffman, S.A.: The Origins of Order: Self-Organization and Selection in Evolution. New York: Oxford University Press (1993)
7. Kauffman, S.A., Levin, S.: Towards a general theory of adaptive walks on rugged landscapes. Theoretical Biology **128** (1987) 11-45
8. Shmulevich,I., Dougherty, E.R., Kim, S., Zhang, W.: Probabilistic Boolean Networks: A rule-based uncertainty model for gene regulatory networks. Bioinformatics **18** (2002) 261-274
9. Shmulevich, I., Dougherty, E.R., Zhang, W.: Gene perturbation and intervention in probabilistic boolean networks. Bioinformatics **18** (2002) 1319-1331
10. Shmulevich, I., Dougherty, E.R., Zhang, W.: Control of stationary behavior in probabilistic boolean networks by means of structural intervention. Biological Systems **10** (2002) 431-446
11. Sierksma, G.: Linear and Interger Programming: Theory and Practice (2002), 2nd ed.

Realization of Biological Data Management by Object Deputy Database System

Zhiyong Peng, Yuan Shi, and Boxuan Zhai

State Key Laboratory of Software Engineering, Computer School
Information Management School, Wuhan University
Wuhan Hubei 430072, China
peng@whu.edu.cn

Abstract. Traditional database system is not suitable for biological data management. In this paper, we discuss how to manage a large amount of complex biological data by an object deputy database system which can provide rich semantics and enough flexibility. In our system, the flexible inheritance avoids a lot of data redundancy. Our cross class query mechanism allows users to find the more related data based on complex relationships. In addition, the schema evolution of biological data and their annotation can be easily supported along with biological knowledge accumulation. The experiment shows our approach is feasible and more efficient than the traditional ones.

1 Introduction

Discovery of sequence of human genome leads to an explosion in biological data. In order to utilize biological data for large scale analysis, it is very necessary to organize and manage data as well as provides convenient data retrieval. More and more databases are being developed for such a purpose. Now there are over a thousand different life science databases with contents ranging from gene-property data for different organisms to brain image data for patients with neurological disorders[12]. According to[6], there are mainly three ways to manage biological data using flat files (36% - 40%), relational databases(37%), and object-oriented databases(8%).

Flat files are usually used to store sequence data. It is more like an exchange format rather than storage format. Although the flat file format has good schema, it does perform ineffectively in the later phrase of data retrieval and data mining. Moreover, data that interests users is always returned as a file in a whole, which imposes a burden on users to extract useful information manually.

Relational databases play an important role in the field of molecular biology. Molecular biology databases based on the relational model often have very complex schemas which, in general, are not intuitive. Therefore, they are often difficult to administrate and to query. Especially, when complex relationships are concerned, too many join operations are needed so that their efficiency is not good.

The object-oriented data model is more suitable than the relational model to model molecular biological data since it can support complex data type and

C. Priami et al. (Eds.): Trans. on Comput. Syst. Biol. V, LNBI 4070, pp. 49–67, 2006.

provide rich semantics. Thus it also received great attention in this field, especially in integration and data warehouse. In addition, object-relational facilities are also utilized to remedies the problems of traditional databases. To date, most efforts to manage this data have relied either on commercial off-the-shelf database management systems developed for business data, or on homegrown systems that are neither flexible nor scalable.

In this paper, we proposed realization of biological data management by object deputy database system, which was implemented based on our object deputy model[24][25]. The model has rich semantics so that the complex semantic relationships of biological data are easy to be defined. Its flexible inheritance mechanisms can avoid a lot of data redundancy. Based on the complex semantic relationships, the cross class query mechanism facilitate data retrieval and data mining. The schema evolution of biological data and their annotations are also easily supported.

The remainder of this paper is organized as follows: Section 2 introduces Object Deputy Model. Section 3 discusses how to realize a biological database based on object deputy model. Section 4 evaluates its performance. We will compare the related works in Section 5. Section 6 concludes this paper.

2   Object Deputy Model

The object-oriented data model[7] is considered to be adequate to meet the requirements stemming from complex, high performance data management such as GIS. However, commercial object-oriented databases[18] are still insufficiently flexible and unable to model the many faceted and dynamic nature of real-world entities. To overcome these deficiencies, various extensions of the object-oriented model are proposed, in particular flexibility and advanced modelling capabilities, such as object views[14], roles[26] and migration[10]. We propose a new approach based on the following analysis.

In relational databases[9], flexibility can be achieved since data are uniformly stored as relation tables, which can be divided and combined freely by relational algebra. However, in order to realize high performance applications, the object-oriented data model encapsulates data and methods in terms of objects. It is difficult to divide and combine objects, thus flexibility of conventional object-oriented databases is rather limited.

To allow objects to be divided and combined freely, we extend the conventional object-oriented data model with the concepts of deputy objects and deputy classes. A deputy object is used to extend and customize its source object(s). Its schema is defined by a deputy class which can be derived by an object deputy algebra like relational algebra. An object can have many deputy objects which may represent its many faceted nature. A deputy object can have its own deputy object as well. Many objects can be combined through a single deputy object. Thus, objects can be divided and combined indirectly through their deputy objects so that the flexibility of object-oriented databases can be easily achieved.

In object-oriented databases, real-world entities are represented in term of
`objects`. Objects are identified by system-defined identifiers which are independent of objects' states. An object has attributes which represent properties of a
corresponding real-world entity. The state of an object is represented by its attribute values, which are read and written by basic methods. In addition, there
are general methods that represent behaviour of objects. Objects having the
same attributes and methods are clustered into `classes` which make it possible
to avoid specification and storage of redundant information. Objects and classes
can be defined as follows.

Definition 1. Each object has an identifier, attributes and methods. Schema of
objects with the same attributes and methods is defined by a class which consists
of a name, an extent and a type. The extent of a class is a set of objects belonging
to it, called its instances. The type of a class is definitions of its attributes and
methods. A class named as C is represented as

$$C = \langle O, A, M \rangle$$

1. O is the extent of C, and $o \in O$ is one of instances of C.
2. A is the set of attribute definitions of C. For $(T_a : a) \in A$, a and T_a represent
   name and type of an attribute, respectively. The value of attribute a of object
   o is expressed by $o.a$. For each attribute $T_a : a$, there are two basic method:
   $read(o, a)$ for reading $o.a$ and $write(o, a, v)$ for writing $o.a$ with the new value
   v, expressed as follows.

$$read(o, a) \Rightarrow \uparrow o.a,$$
$$write(o, a, v) \Rightarrow o.a := v$$

   Here, $\Rightarrow$, $\uparrow$ and $:=$ stand for operation invoking, result returning and assign-
   ment, respectively.
3. M is the set of method definitions of C. For $(m : \{T_p : p\}) \in M$, m and
   $\{T_p : p\}$ are method name and a set of parameters, p and T_p represent
   parameter name and type, respectively. Applying method m to object o
   with parameters $\{p\}$ is expressed as follows. For simplification, methods are
   not considered to return results here.

$$apply(o, m, \{p\})$$

Deputy objects are defined as extension and customization of objects. The
schemata of deputy objects are defined by deputy classes that are derived by
creating deputy objects as their instances, generating switching operations for
inheritance of attributes and methods, and adding definitions for their additional
attributes and methods as well as constraints. Note that the switching operation
is peculiar to deputy objects. It is used to switch an operation request from
a deputy object to its source object. Thus, the source object can execute the
operation for the deputy object. From the user's view, it seems as if the deputy
object executed the operation. During the switching process, the operation request can be changed into the form suitable for the source object so that the

capability of an object can be customized for different application situation. In general, deputy objects and deputy classes can be defined as follows.

Definition 2. A deputy object is generated from object(s) or other deputy object(s). The latter is called source object(s) of the former. A deputy object must inherit some attributes/methods from its source object(s) in order to avoid generating nonsense deputy objects. The schema of deputy objects with the same properties is defined by a deputy class, which includes a name, extent and type. Deputy classes are derived from classes of source objects called source classes. In general, let $C^s = \langle O^s, A^s, M^s \rangle$ be a source class. Its deputy class C^d is defined as

$$C^d = \langle O^d, A^d \cup A_+^d, M^d \cup M_+^d \rangle$$

1. O^d is the extent of C^d. There are the following three cases.
   (a) $O^d = \{o^d \,|o^d \to o^s, sp(o^s) == true, o^s \in O^s\}$, where $o^d \to o^s$ represents that o^d is the deputy object of a single object o^s, and sp represents selection predicate.
   (b) $O^d = \{o^d|o^d \to (... \times o^s \times ...), cp(... \times o^s \times ...) == true, o^s \in O^s\}$, where $o^d \to (... \times o^s \times ...)$ represents that o^d is the deputy object of $(... \times o^s \times ...)$ that are several objects with possible different types, and cp represents combination predicate.
   (c) $O^d = \{o^d|o^d \to \{o^s\}, gp(\{o^s\}) == true, o^s \in O^s\}$, where $o^d \to \{o^s\}$ represents that o^d is the deputy object of $\{o^s\}$ that is a set of objects with the same type, and gp represents grouping predicate.
2. $A^d \cup A_+^d$ is the set of attribute definitions of C^d.
   (a) $(T_{a^d} : a^d) \in A^d$ is the attribute inherited from $(T_{a^s} : a^s) \in A^s$ of C^s, of which switching operations are defined as

   $$read(o^d, a^d) \Rightarrow\uparrow f_{T_{a^s} \mapsto T_{a^d}}(read(o^s, a^s)),$$
   $$write(o^d, a^d, v^d) \Rightarrow write(o^s, a^s, f_{T_{a^d} \mapsto T_{a^s}}(v^d))$$

   (b) $(T_{a_+^d} : a_+^d) \in A_+^d$ is the additional attribute of C^d, of which basic methods are defined as
   $$read(o^d, a_+^d) \Rightarrow\uparrow o^d.a_+^d,$$
   $$write(o^d, a_+^d, v_+^d) \Rightarrow o^d.a_+^d := v_+^d$$

3. $M^d \cup M_+^d$ is the set of method definitions of C^d.
   (a) $(m^d : \{T_{p^d} : p^d\}) \in M^d$ is the method inherited from $(m^s : \{T_{p^s} : p^s\}) \in M^s$ of C^s, which is applied through switching operation as

   $$apply(o^d, m^d, \{p^d\}) \Rightarrow apply(o^s, m^s, \{f_{T_{p^d} \mapsto T_{p^s}}(p^d)\})$$

   (b) $(m_+^d : \{T_{p_+^d} : p_+^d\}) \in M_+^d$ is the additional method of C^d, which is applied as

   $$apply(o^d, m_+^d, \{p_+^d\})$$

The object deputy model provides an object deputy algebra for deputy class derivation, which contains the following six operations.

1. The **Select** operation is used to derive a deputy class of which instances are the deputy objects of the instances of a source class that are selected according to a selection predicate.
2. The **Project** operation is used to derive a deputy class which only inherits part of attributes and methods of a source class.
3. The **Extend** operation is used to derive a deputy class of which instances are extended with additional attributes and methods that cannot be derived from a source class.
4. The **Union** operation is used to derive a deputy class of which extent consists of deputy objects of instances of more than one source class.
5. The **Join** operation is used to derive a deputy class of which instances are deputy objects for aggregating instances of source classes according to a combination predicate.
6. The **Grouping** operation is used to derive a deputy class of which instances are deputy objects for grouping instances of a source class according to a grouping predicate.

There are semantic constraints that are defined as predicates of deputy classes. They are expressed as boolean expression, such as aStudent.age > 15 being used as selection predicate. The selection predicate determines existence of a deputy object according to the state of its single source object. Only when the source object satisfies some special condition, its deputy object can exist. The combination and grouping predicates define the existence conditions between a deputy object and several source objects. In order to enforce these semantic constraints, data update propagations between deputy objects and their source objects need to be supported.

We have designed three procedures for data update propagation, which are caused by addition of an object, deletion of an object and modification of one attribute value of an object, respectively. They are realized based on bilateral links between objects/classes and their deputy objects/classes as well as various semantic constraints. That is, when the above basic update operations occur on some classes, their deputy classes will be examined so that the updates on deputy classes can be caused in order to maintain semantic constraints defined by deputy classes. The update procedures may be invoked recursively, since deputy classes may have their own deputy classes.

3   Biological Data Management

3.1   Biological Data Modelling

We will take molecular biological data management[23] as an example, which involves gene, protein, chromosome, genome, mutation and etc. These biological objects have complex relationships.

For instance, the complete genome of a species consists of a collection of chromosomes, each of which can be considered to be a long sequence of DNA, which in turn consists of a mass of (potentially overlapping) chromosome fragments. These fragments are either Transcribed regions or Non-Transcribed regions of DNA. A gene is a segment of the chromosome that is transcribed into RNA which includes transcribed regions along with flanking non-transcribed sequence/region. There is also possibility that two adjacent genes may overlap, where a part of the coding sequence, of the previous one, acts as the promoter for the following one. A gene may exist in different chromosomes, if segments with the same function are found on them. Each modification to a gene is represented as a mutant, which can be classified according to the mutation size, such as small change involving not more than 20 bps and gross change having larger modification.

From another aspect, one transcribed chromosome fragment encodes one or more proteins under different environments. In their multi-multi relationships, mRNA (message RNA) plays a unique part in transporting the coding information from DNA to a protein. Further, proteins, which belong to a certain protein family, can interact with each other. The non-transcribed regions are either regulators, which control the expression of genes, or chromosomal elements, which include the centromere, the telomeres and the origins of replication (ORI). The relationships among these concepts can be shown in Fig. 1.

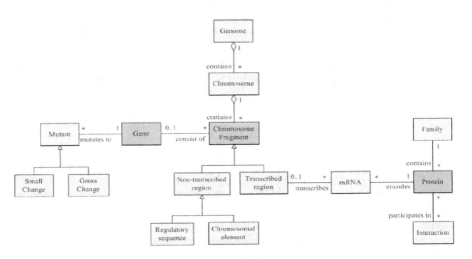

Fig. 1. The Conceptual Model for a Biological Data Database

The above example shows the molecular biological data is very complex. Firstly, the core data, such as the genome, gene and protein, themselves are very complex. In addition, it is very hard to describe complex relationship between these information, which increases the complexity of building biology database. Many researchers have proposed various conceptual models to define the extremely complex biological data[17][23]. However, these conceptual models are more or less poorly supported by traditional databases.

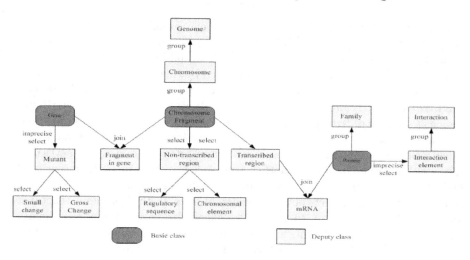

Fig. 2. The Biological Data Schema in Object Deputy Database

Using object deputy model, the complex relationships of the molecular biological data can be defined easily. The multi-to-one relationship such as "contains" can be defined by the group deputy class. The inheritance of specialization corresponds to the select deputy class. And the complex aggregation relationship can be represented by the join deputy class. Moreover, we can use the "imprecise" select deputy class to create multiple duplications of one entity in one deputy class. In this way, the biological data schema can be defined as shown in Fig. 2, where arrows denote the direction from the source class to the deputy class and the words beside arrows show what kinds of deputy classes are defined.

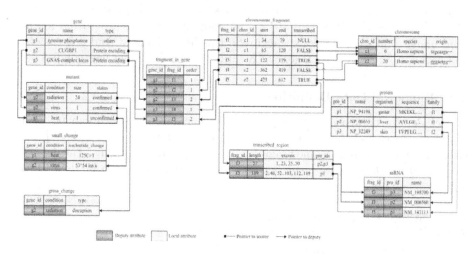

Fig. 3. The Biological Data Structure in Object Deputy Database

Objects and deputy objects are instances of classes and deputy classes, respectively, which are connected by bilateral links as shown in 3. There are less data redundancy since attribute values of objects are inherited by their deputy objects. For example, $gene\_id$ of $gene$ is inherited by their mutants; The attributes $start$ and end of the class $chromosome\ fragmant$ are inherited by the deputy class $transcribed\ region$ as length. Deputy objects have additional attributes such as $locus\_tag$ of $transcribed\ region$ and $condition$ of $mutant$. The inherited attributes of deputy objects do not take up physical space but their values are calculated temporally during data access. This strategy, in one aspect, saves disk space without redundancy, in another aspect effectively ensures the integrity of biological data. Besides, inheritance through switching operations also avoids engendering any data conflicts between related objects. With these semi-structured features, which is difficult to achieve in other databases, our object deputy database can store the growing entities well without data redundancy. In detail, users can store the basic core properties of objects in a source class, and create various deputy classes for special subsets of objects. Thus, deputy objects of different kinds can reuse properties of interest from their source objects through switching operations and extend relevant additional properties for particular applications when more detailed information are discovered. The typical examples are the deputy classes $transcribed\_region$ and $non-transcribed\_region$, which represent two different subsets of the $chromosome\_fragment$ objects. In practice, when a novel fragment is discovered and only its basic information is known, it will be firstly stored in class $chromosome\_fragment$. After it is fully researched and more properties of it are available, the corresponding deputy objects may be created in $transcribed\_region$ or $non-transcribed\_region$, which can store more information of it. A similar situation is in imprecise deputy classes, where the multiple deputy objects share the common data of their source object and have different values of the extended attributes. For instance, a "gene" object can have several deputy objects in "mutant" class resulting from mutations under different conditions which reflect reality in biology. And it is easy to add new mutant objects for a gene to describe its mutant under new conditions.

3.2   Biological Data Definition

Object deputy database provides a SQL-like language[28] which can be used to define classes and their deputy classes. In this example, there are three basic classes, "Gene", "Chromosome Fragment" and "Protein", from which the deputy classes $transcribed\_region$, $small\_change$ and $chromosome$ can be defined as follows.

Create SELECT DeputyClass $transcribed\_region$
SELECT $frag\_id, (end-start)\ AS\ length$
FROM $chromosome\_fragment$
WHERE $trascribed = TRUE$
EXTEND $(extons\ VARCHAR,\ pro\_ids\ ARRAY)$;

The selection deputy class *transcribed_region* has attributes *frag_id* and *length* inherited from the souce class *chromosome_fragment* through switching operations such as $(end - start)$ *AS length*. The boolean expression *transcribed* $= TRUE$ is used to define the selection predicate, where the *transcribed* is used to store the experiment result that whether the fragment can be transcribed. As a transcribed region, it may be divided into discontinuous exons and can encodes one or more proteins. We use EXTEND clause to define the additional attributes "exons" for *transcribed_region* to store the start and end positions of exons and *pro_ids* to record the proteins it encodes.

Based on the extended attribute *pro_ids* in *transcribed_region* , we can define the join deputy class *mRNA* from *transcribed_region* and *protein* using the following statement.

Create JOIN DeputyClass *mRNA*
SELECT *frag_id, pro_id*
FROM *transcribed_region, protein*
WHERE *protein.pro_id* IN *transcribed_region.pro_ids*
EXTEND (*name VARCHAR*);

When it is impossible or at least difficult to define concrete selection predicates, users can choose to create "imprecise" deputy classes. Since it is hard to abstract precise mutation regularity from the gene's attributes, the "mutant" can be defined as an imprecise deputy class.

Create Imprecise SELECT DeputyClass *mutant*
SELECT *gene_id*
FROM *gene*
EXTEND (*condition VARCHAR, size INTEGER, status VARCHAR*);

The propagation module will not create deputy objects for the imprecise deputy class automatically. We defined a special syntax to create imprecise deputy objects manually as follows:

ADD ANY INTO *mutant* FROM *gene* WHERE *gene_id* = *g2*
WITH (*condition, size, status*) VALUES (*'radiation'*, 24, *'confirmed'*);

The above statement creates a deputy object in mutant for the gene object "g2", where the WITH clause declares the values of the extended attributes. The *mutant* has the two deputy classes *small_change* and *gross_change*. A small change is a modification involving not more than 20 nucleotides, and may be either substitution, insertion or deletion. A gross change is a larger mutant to a gene, which could be the disruption of the gene, the complete replacement of the whole gene with another gene that can be expressed, or the complete deletion of the entire gene, and it can also be a partial replacement or deletion of a length of sequence within the gene. In our model, these two deputy classes inherit the common attributes, *gene_id* and *condition*. And they respectively hold the deputy constrains of *mutant.size* $<= 20$ and *mutant.size* > 20. The *small_change* has the additional attribute *nucleotide_change* to describe the detail of how the nucleotides are changed, and the *gross_change* has the additional attribute *type* to indicate what type the gross change is.

Create GROUP DeputyClass *chromosome*
SELECT *chro_id*
FROM *chromosome_fragment*
GROUP BY *chro_id*
EXTEND (*number INTEGER, species VARCHAR, origin TEXT*);

The group deputy class *chromosome* is created by grouping chromosome fragments according to their *chro_id*. It is extended with three attributes *number*, *species*, and *origin*, which respectively represent the number of the chromosome in its genome, the species the chromosome belongs to, and the whole DNA sequence of the chromosome.

3.3   Biological Data Evolution

The important feature of the biological data is its semi-structure and having some unknown attributes and relationships. People can find these attributes and relationships gradually by doing various experiments. Therefore, the biological data schema should be allowed to be modified easily. However, the current DBMSs, even extended with object-relational facilities, support poorly schema evolution.

In our object deputy database, one can define a deputy class with additional attributes if one wants to add the new found attributes to the biological objects. The data of the source class can be fully reused while avoiding the trouble of modifying the schema of the existing classes. It is also convenient to adjust the schema of classes (such as change the attribute names and/or types; conceal private properties) through their deputy classes to satisfy the interface requirements of special applications. For example, we can create a new deputy class *function* from the class *gene* as a supplement for newly-found research result of the functions of particular genes.

In classical databases, defining the new relationships usually means modifying the schema of the existing tables/classes by adding new foreign keys or associations. Whereas in object deputy database, this can be easily realized by defining a join deputy class on related classes to link related objects and describe the specific properties of the relationship. For example, we can create a join deputy class *mutant_protein* from the classes *mutant* and *protein* to describe the affections of a mutant on the corresponding proteins.

3.4   Biological Data Annotation

Currently, more and more focus has been shift to the knowledge about molecular-biological objects, such as, genes, proteins, intra- and inter-cellular pathways, etc., which is typically encoded by a large variety of data commonly called annotations. Such annotations are continuously collected, created, and made available in numerous public data sources.

Thus, freely creating various personal annotations and effectively organizing them is a special requirement in biology databases. But the miscellaneousness of various annotations becomes a great obstacle of organizing this special data. In one aspect, a same object or phenomenon may be annotated by different scientists with conclusions coming from their own experiments under various environments. At another aspect, the same scientist may have his/her particular opinions on objects of different types. In addition, how to efficiently retrieve annotations of interest from such an intricate data set remains a problem.

In our object deputy database, the flexibility of imprecise deputy class meets the requirement of personal annotation very well. Consider, for example, one can create imprecise deputy classes *disease_annotation* and *sequence_annotation* from *protein* to annotate proteins from different aspects. Thus, any scientist can freely add their annotations of these two kinds at any time. As shown in Fig. 4, different people (Prof. Zhai and Prof. Shi) can add their annotations to the same object (object p2) of the same kind (disease); the same people (Prof. Zhai) can annotate the same object differently.

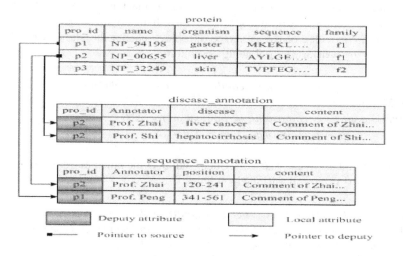

Fig. 4. The Biological Data Annotation

3.5   Biological Data Query

A great difference between the biological data management and classical commercial data applications is the substantial number of data types visited in a single biological data analysis task, which involves a wide range of objects. Another problem is that most of biology database users are biologists that are not familiar with database schemas and even the query languages such as SQL. Therefore, the popular biology databases in general have form-based query and/or browsing interfaces, which is convenient for simple queries but much limited in complex data access.

With a powerful navigation mechanism, our object deputy database supports the cross-classes query efficiently and conveniently. Since related objects in our object deputy database are connected with bi-directional pointers, it is easy to navigate from one object to any of its related objects via these pointers. We use the symbol " $\rightarrow$ " to represent the navigation between classes.

Definition 3. Let C^b be the class from which the navigation begins and C^t be the target class. For any object $o^b \in C^b$, the expression $C^b \rightarrow C^t$ will return all its connected objects $\{o^t\} \in C^t$. If no target object could be found, the result will be empty. The expression "$C^b \rightarrow C^t$" is called a "path expression".

An interesting feature of the path expression in our object deputy database is that only the "start" and "target" classes should be explicitly written in query. Since the deputy relationship forms a directed graph, one start-to-target class pair determines an unique path. In this way, the database users can be greatly benefited from getting rid of understanding perplexing data schema and expressing complex query. With these features, path expression makes the query not only simple and easier to understand but also more efficient to execute. For example (see Fig. 3), an expert whose research is focused on human liver cancer may want to get the cross-change mutation information of genes that is located in human chromosome and encodes protein in liver organism. Even he/she does not have a clear idea of the concrete database schema, he/she can quickly get the answer by writing the following statement:

SELECT $gene \rightarrow cross\_chage.(condition, nucleotide\_change)$
FROM $gene$
WHERE $gene \rightarrow chromosome =' homosapiens'$ AND
 $\quad gene \rightarrow protein.organism =' liver'$

In relational database, the same query will become much more complex, which involves a lot of join operations as follows.

SELECT $mutant.condition, cross\_change.nucleotide$
FROM $gene\ g, gene\_frags\ gf, fragment\ f, chromosome\ c, transcribed\ t,$
 $\quad mRNA\ mr, protein\ p, mutant\ m, cross\_change\ c$
WHERE $g.gene\_id = m.gene\_id$ AND $m.mutant\_id = c.mutant\_id$ AND
 $\quad g.gene\_id = c.gene\_id$ AND $g.gene\_id = gf.gene\_id$ AND
 $\quad gf.fragment\_id = f.fragment\_id$ AND
 $\quad f.chromosome\_id = c.chromosome\_id$ AND
 $\quad f.chromosome\_id = c.chromosome\_id$ AND
 $\quad c.chromosome\_name =' homosapiens'$ AND
 $\quad f.fragment\_id = t.fragment\_id$ AND
 $\quad mr.trancribed\_id = t.trancribed\_id$ AND
 $\quad mr.protein\_id = p.protein\_id$ AND $p.organism =' liver';$

Note that the path expression of our object deputy database is not the same as that of the object-oriented databases, which use a class as its attribute domain to navigate from an object to its sub-objects. The attribute-based path expression degrades the interactional relationship to be a unilateral one. Thus, the object paths of this kind are single-directed and difficult to be extended. Further more, no matter in relational databases or object-oriented databases, to execute a path query, all the tables/classes and indexes involved in the path must be visited. Whereas, our object deputy database only visits the start class and target class as well as the object mapping table, which stores all the bi-directional pointers in database.

4   Efficiency Evaluation

We have conducted several experiments to evaluate the feasibility and efficiency of biological data query in our object deputy database, especially for a single query that involves complex relationships. Our main purpose is to compare the performance between the object relational database PostgreSQL and our object deputy database Totem with equal scale of data. The test is simultaneously carried out on two PC of 2G Hz Pentium IV processor, 256 M internal RAM, 80G total disk space and Redhat Linux 9.0 OS. On the two servers, we separately installed PostgreSQL7.2, a famous open source object relational database system, and TOTEM1.0, an object deputy database system developed by the authors. The schemas of biological data in these two databases are respectively shown in Fig 5 and Fig 2. The data scale is arranged as gene number: 10 thousands, protein number: 50 thousands, chromosomes fragment: 100 thousands.

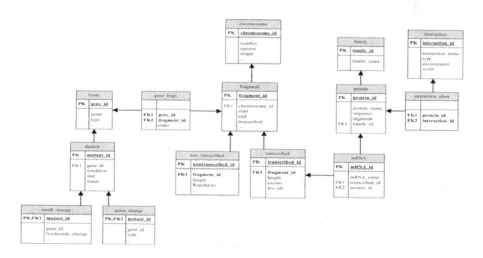

Fig. 5. The Biological Data Schema in Relational Database

Query 1 In PostgreSQL	SELECT i.*, f.* FROM interaction i, interaction_elem e, protein p, family f WHERE interaction_name = input AND i.interaction_id = e.interaction_id AND e.protein_id = p.protein_id AND p.family_id = f.family_id ;
Query 1 In TOTEM	SELECT interaction.*, interaction-family.* FROM interaction WHERE interaction_name = input ;
Query 2 In PostgreSQL	SELECT g.*, m.*, e.* from gene g, mutant m, gross_change gr WHERE g.gene_name = input AND g.gene_id = m.gene_id AND m.mutant_id = gr.mutant_id ; SELECT g.*, m.*, s.* from gene g, mutant m, small_change s WHERE g.gene_name = input AND g.gene_id = m.gene_id AND m.mutant_id = s.mutant_id ;
Query 2 In TOTEM	SELECT gene.*, gene-gross_change.* FROM gene WHERE gene.gene_name = input ; SELECT gene.*, gene-small_change.* FROM gene WHERE gene.gene_name = input ;
Query 3 In PostgreSQL	SELECT g.*, f.*, t.* FROM gene g, gene_frags gf, fragment f, transcribed t WHERE g.gene_name = input AND g.gene_id=gf.gene_id AND gf.fragment_id = f.fragment_id AND gf.fragment_id=t.fragment_id ; SELECT g.*, f.*, n.* FROM gene g, gene_frags gf, fragment f, non_transcribed n WHERE g.gene_name = input AND g.gene_id=gf.gene_id AND gf.fragment_id = f.fragment_id AND gf.fragment_id=n.fragment_id;
Query 3 In TOTEM	SELECT gene.*, gene-transcribed.* FROM gene WHERE gene.gene_name = input ; SELECT gene.*, gene-non_transcribed.* FROM gene WHERE gene.gene_name = input ;
Query 4 In PostgreSQL	SELECT g.*, m.*, p.* FROM gene g, gene_frags gf, transcribed t, mRNA m, protein p WHERE g.gene_name = input AND g.gene_id = gf.gene_id AND gf.fragment_id = t.fragment_id AND t.transcribed_id = m.transcribed_id AND m.protein_id = p.protein_id;
Query 4 In TOTEM	SELECT gene.*, gene-mRNA.*, gene-protein.* FROM gene WHERE g.gene_name = input ;

Fig. 6. The Typical Biological Data Query

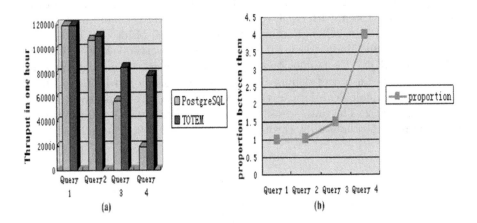

Fig. 7. The Experiment Result

Based on the model in Fig. 1, we designed four typical biological data query examples:

1) Given an interaction, find the related protein family. It will be useful in summarizing the knowledge of family-interaction.
2) Given a gene name, find its mutant and the detail information of two mutant types. This is a common query for those who focuses on the alleles of a gene.
3) Given a gene name, find all its fragment information, including both the transcribed and non-transcribed fragments. This query is useful to get the general picture of a gene.

4) Given a gene name, find the information of proteins that it encodes as well as that of the involved mRNA. This query is useful when someone wants to know how a gene encodes its related proteins.

The statements of these queries are represented in fig. 6.

In fig.7 (a), x-axis denotes 4 queries, and y-axis records the number of successful transactions executed in an hour of each query. From the experiment results shown in this graph, we can safely get the conclusion that Totem works better than PostgreSQL in almost every query. In fig.7 (b), we can see that, although the two databases bears similar efficiency on executing simple queries, TOTEM shows huge predominance when the query is complex and involves a lot of classes/relations.

5   Related Works

Till now, a variety of biological or related databases have been developed, ranging from simple sequence repositories, which store data with little or no manual invention in the creation of the records, to expertly create universal databases that cover all species and the original sequence data together with the manual addition of further information.

Flatfile Databases: In the early days when rare database management technique was used in the field of molecular biology, data is always organized into indexed ASCII text files, so called "flat files", which in general have no explicit data models and seem to have no common semantic structure. The most typical examples that use flat files are the International Nucleotide Sequence Database Collaboration: Genebank[5], EMBL[3] and DDBJ[27]. Genebank is a well-known nucleotide sequence database built by National Center for Biotechnology Information (NCBI). Following the idea of flat files, each entry in Genebank is organized as a file that has a concise description of the sequence, the scientific name and the taxonomy of the source organ, bibliographic reference and a table of sequence. To identify the entry in the file, Genebank staffs manually assign an accession number to a submitted sequence within 2 work days, and do so at a rate of several hundreds per day. EMBL is a European database to distribute most nucleotide sequence data and support bibliographical and biological data. It designed documents to describe the structure of entries in the EMBL database which is split into 2 parts: "common" schema that represents the elements shared by all EMBL entries and "main" schema that incorporates the common ones. DDBJ is another DNA sequence database that stores the data submitted somewhat reflects the trend in biological research in Japan. In the purpose of exchanging data freely and conveniently, standard flat file format was defined. In specific, all the three databases exchange data with a newly improvement of common XML format. Besides the databases mentioned, Swissprot[2] is another typical flat file system established in 1986 to store annotated protein sequence. Sequence entries are composed of different line types, each with their own format close to that of EMBL.

Relational Databases: In the mid of nineties, the relational model was brought into this field. Among these relational databases, ALFRED focuses on allele frequencies in diverse anthropologically defined populations. Initially, using Microsoft Access, it now has been implemented using Oracle version 8.1.7.4. For its detailed table information and relationships, refer to the document in [1]. Another example is Biological Magnetic Resonance Data Bank (BMRB), which is designed on Informix database as a repository for data from NMR spectroscopy on proteins, peptides, and nucleic acids. Besides, PDBase[11] was built in Oracle to record the expression of macromolecular structure data.

Object Databases: Not satisfied with the relational model, many institutions resort to object-oriented database as a solution. aMaze[15] is an example that embodies general rules for associating individual biological entities and interactions into large complex networks of cellular processes. It designed three main classes in the databases, respectively, BiochemicalEntity, Interaction and Process. GeneDB[21], encoded with three main objects, Regions, Observations an Annotation, is an open source genome annotation system that is based on Linux computer operating system. SPINS[20] is a data dictionary and object-oriented database for archiving protein NMR spectra. In the concrete process of implementation, these databases have the common characteristic that their data is manipulated as objects, but stored in tables using a relational DBMS engine.

Object-Relational Model: Other than the above models, the object-relational database is also a feasible solution that is applied by modelers. PIR-PSD[4], iProClass[16] and PIR-NREF databases are examples that have been implemented in Oracle 8i object-relational database system on Unix server. As discussed before, although these mentioned databases have already been widely used by biologists, they either remain using the primal file system or mature database products that are developed for other applications, none of which adapts the biological data well. Meanwhile, two unconventional databases, in particular, AceDB and OPM, are designed specifically for molecular biology data management. AceDB[22] was initially developed for a database called "A C.elegans Data Base (ACeDB)". Another example that adopts this strategy is a genetic database called SacchDB. AceDB resembles an object database management system while supporting neither class hierarchies, nor inheritance. An AceDB object has a set of attributes that are objects or atomic values such as numbers or strings. AceDB objects are represented as trees where the (named) nodes are objects or atomic values and arcs express the attribute relationships. Like our object deputy model, AceDB has the advantages of accommodation to irregular objects and good schema extension for classes. However, it possesses less flexibility on describing the relationships between objects and classes. Firstly, AceDB follows the unilateral object link strategy in object-oriented databases, which can not ensure efficient object navigation. Secondly, abandoning inheritance leads to the inability of representing class hierarchy and sharing data. Finally, AceDB is refuted with many deficiencies at the same time, such as inability of scaling to large projects, rather homegrown technology, lack of modern concurrency tools and limit publications support, all of which make it not so attractive.

Object-Protocol Model: Dissatisfied with the inadequacy of traditional model ability, Object-Protocol Model (OPM)[8] is developed for modelling complex scientific experiments. Together with a suit of OPM-based data management tools, it can also represent heterogeneous databases in a uniform, abstract, and consistent way with its specific query language. Its open source distribution, coupled with its ready-made graphical user interface and Perl and Java APIs makes it an attractive choice for groups with modest informatics resources, or for those who wish to avoid the slow development times associated with general-purpose database engines. There are several OPM-based public molecular biology databases, such as the Genome Data Base (GDB)[19] at Johns Hopkins School of Medicine, the Genome Sequence Data Base (GSDB)[13] at the National Center for Genome Resource and so on. To some extent, the OPM model bears many similarities with our object deputy model e.g., with OPM views, users can customize data and schema of classes just as deputy class in object deputy model; the introduction of "protocol" and "input/output attributes", which enables OPM to describe complex experimental relationship between classes and objects, only supports part of the function of bilateral pointer mechanism in object deputy model; like the update propagation in object deputy model, OPM also supports defining various rules to maintain the dependence consistency between objects. However, OPM is mainly designed for describing the scientific experiment and has no special database kernel program.

6   Conclusion

Data modelling is really a difficult work in the field of the molecular biology, the difficulty comes not only from the intricate relationship between complex data, but also from the inability in expression ability of traditional data models. By analyzing current classical data model and their respective advantages and disadvantages, we found our object deputy model is suitable for biological data management due to its rich semantics and great flexibility. The experiment shows our object deputy database has better efficiency than the traditional ones. In the future, we are going to improve our databases with more optimization technologies.

Acknowledgment

This work was supported by National Natural Science Foundation of China (60573095), Program for New Century Excellent Talents in University of China (NCET-04-0675), Research Funding for Doctoral Program of Higher Education (20050486024) and State Key Lab of Software Engineering (Wuhan University, China) under grant:SKLSE05-01. We would like to thank biology laboratories for the special knowledge they provided, especially to Mrs Yiwen Bu for her helpful assistance during our modelling design.

References

1. An allele frequency database for diverse populations and dna polymorphisms data structure in alfred. The Board of Regents of the University of Wisconsin System.
2. A. Bairoch and R. Apweiler. The swiss-prot protein sequence database and its supplement trembl in 2000. In *Nucleic Acids Res. Vol. 28, No. 1*, pages 45–48. Oxford University Press, 2000.
3. W. Baker, A. van den Broek, E. Camon, P. Hingamp, P. Sterk, G. Stoesser, and M. Tuli. *The EMBL Nucleotide Sequence Database*. 2000.
4. W. C. Barker, J. S. Garavelli, H. Huang, P. B. McGarvey, B. C. Orcutt, and etc. The protein information resource(pir). In *Nucleic Acids research 2000, vol. 28, no. 1*, pages 41–44. Oxford University Press, 2000.
5. D. Benson, I. Karsch-Mizrachi, D. Lipman, J. Ostell, B. Rapp, and D. Wheeler. *Genebank*. 2000.
6. F. Bry and P. Kroger. A computational biology database digest: Data, data analysis, and data management. In *Distributed and Parallel Databases*, pages 7–42. Kluwer Academic Publishers Hingham, MA, USA, January 2003.
7. R. Cattell. *The Object Database Standard:ODMG-93 (Release 1.1)*. Morgan Kaufmann, San Francisco, CA, 1994.
8. I.-M. Chen and V. Markowitz. an overview of the object protocol model(opm) and the opm data management tools. In *Information Systems(1995)*, 398-418 1995.
9. E. Codd. A relational model of data for large shared data banks. *Communication of the ACM*, 13(6):377–387, 1970.
10. M. E.El-Sharkawi and Y. Kambayashi. Object migration mechanisms to support object-oriented databases. In *Databases: Theory, Design, and Application*, pages 73–91, 1991.
11. D. S. Greer. Pdbase: A relational database expression of macromolecular structure data.
12. A. Gupta. Life science research and data management c what can they give each other? In *ACM SIGMOD 2004 Electronic Proceedings, Paris, France*, pages 12–14. ACM, June 2004.
13. C. Harger, M. Skupski, J. Bingham, A. Farmer, and S. Hoisie. *The Genome Sequence DataBase(GSDB): improving data quality and data access*. 1998.
14. S. Heiler and S. Zdonick. Object views: Extending the vision. In *Proc. of IEEE Sixth Int. Conf. On Data Engineering*, pages 86–93, 1990.
15. M. G. Hicks and C. Kettner. amaze:a database of molecular function,interactions and biochemical processes. In *Proceedings of the Beilstein-Institut Workshop*, May 2002.
16. H. Huang, W. C. Barker, Y. Chen, and C. H. Wu. iproclass: an integrated database of protein family classification, function and structure information. In *Nucleic Acids Research 2003, vol. 31*, page 390C392. Oxford University Press, 2003.
17. C. M. M. Keet. Biological data and conceptual modelling methods. In *Journal of Conceptual Modeling*, October 2003.
18. W. Kim. Object-oriented database systems: Promises, reality, and future. In *Proc. of the 19th VLDB*, pages 676–687, 1993.
19. S. Letovsky, R. Cottingham, C. Porter, and P. Li. *GDB: the Human Genome Database*. 1998.
20. B. MC, M. HN, S. G, and M. GT. Spins: standardized protein nmr storage. a data dictionary and object-oriented relational database for archiving protein nmr spectra. October 2002.

21. Meyer and F. et al. *GenDBan open source genome annotation system for prokaryote genomes*. 2003.

22. J. T. Mieg, D. T. Mieg, and L. Stein. Acedb: A genome database management system. In *Computing in Science and Engineering, vol. 1, no. 3*, pages 44–52, May/June 1999.

23. N. Paton, S. Khan, A. Hayes, and et al. conceptual modeling of genomic information. In *Bioinformatics*, pages 548–557, June 2000.

24. Z. Peng. *An Object Deputy Model for Advanced Database Application*. PhD Thesis, Kyoto University, Japan, 1994.

25. Z. Peng and Y. Kambayashi. Deputy mechanisms for object-oriented databases. In *Proceedings of the IEEE 11th Int. Conf. of Data Engineering*, pages 333–340, 1995.

26. J. Richardson and P. Schwariz. Aspects:extending objects to support multiple, independent roles. In *Proc. of the Int. Conference on Management of Data*, pages 298–307, 1991.

27. Y. Tateno, S. Miyazaki, M. Ota, H. Sugawara, and T. Gojobori. *DNA Data Bank of Japan (DDBJ) in Collaboration with Mass Sequencing Teams*. 2000.

28. B. Zhai, Y. Shi, and Z. Peng. Object deputy database language. In *The Fourth International Conference on Creating, Connecting and Collaborating through Computing (C5 2006)*, January 2006.

GeneTUC, GENIA and Google: Natural Language Understanding in Molecular Biology Literature

Rune Sætre[1], Harald Søvik[1], Tore Amble[1], and Yoshimasa Tsuruoka[2]

[1] Department of Computer and Information Science,
Norwegian University of Science and Technology,
Sem Sælandsv. 7-9, NO-7491 Trondheim, Norway
Rune.Satre@idi.ntnu.no
http://www.idi.ntnu.no/~satre

[2] Department of Computer Science, University of Tokyo,
Hongo 7-3-1, Bunkyo-ku, Tokyo 113-0033, Japan

Abstract. With the increasing amount of biomedical literature, there is a need for automatic extraction of information to support biomedical researchers. GeneTUC has been developed to be able to read biological texts and answer questions about them afterwards. The knowledge base of the system is constructed by parsing MEDLINE abstracts or other online text strings retrieved by the Google API. When the system encounters words that are not in the dictionary, the Google API can be used to automatically determine the semantic class of the word and add it to the dictionary. The performance of the GeneTUC parser was tested and compared to the manually tagged GENIA corpus with EvalB, giving bracketing precision and recall scores of 70,6% and 53,9% respectively. GeneTUC was able to parse 60,2% of the sentences, and the POS-tagging accuracy was 86.0%. This is not as high as the best taggers and parsers available, but GeneTUC is also capable of doing deep reasoning, like anaphora resolution and question answering, which is not a part of the state-of-the-art parsers.

Keywords: Biomedical Literature Data Mining, Google API, GENIA.

1   Introduction

Modern research is presenting more new and exciting results than ever before, and it is gradually becoming impossible for the human reader to stay up-to-date in the sea of information. This is especially true in the Medical and Molecular Biology domains, where the MEDLINE database of publications is increasing with almost 2000 new entries every day. To help researchers find the information they are searching for in an efficient manner, automatic Information Extraction (IE) is needed. This paper describes a system that is using Natural Language Processing (NLP) in order to automatically read the abstracts of research papers, and later answer questions posed in English about the abstracts.

C. Priami et al. (Eds.): Trans. on Comput. Syst. Biol. V, LNBI 4070, pp. 68–82, 2006.
© Springer-Verlag Berlin Heidelberg 2006

1.1 Information Extraction (IE) in Biology

The large and rapidly growing amounts of biomedical literature demands a more *automatic* extraction process than previously. Current extraction approaches have provided promising results, but they are not sufficiently accurate and scalable. A survey describing different approaches within the *information extraction field* is presented in [6], and a more recent "IE in Biology" survey is given in [15]. In the biomedical domain, IE approaches range from simple automatic methods to more sophisticated but also more manual methods. Some good examples are: Learning relationships between proteins/genes based on co-occurrences in MEDLINE abstracts [9], *manually* developed IE rules [24], protein name classifiers trained on *manually* annotated training corpora [1], and classifiers trained on *automatically* annotated training corpora [20].

A new emerging approach to medical IE is the heavy use of corpora. The workload can then be shifted from the extremely time consuming manual grammar construction to the somewhat easier and more teamwork oriented corpus/treebank building [12]. This means that the information acquisition bottleneck can be overcome, while still reaching state-of-the-art coverage scores (around 70-80 percent). In this chapter a corpus is used, namely the GENIA Tree Bank (GTB) corpus [19], first to train and then later to test how well the GeneTUC parser performs compared to other available parsers in this domain.

1.2 GeneTUC

The application that we want to improve and test, by incorporating alternative sources of information, is called GeneTUC. TUC is short for "The Understanding Computer", and is a system that is under continuous development at the Norwegian University of Science and Technology. Section 3 will explain in more detail how TUC, and especially GeneTUC, works.

1.3 Chapter Structure

The rest of this chapter is organized as follows. Section 2 describes the materials and programs that were used, section 3 explains in detail how GeneTUC works, section 4 presents our approach, section 5 presents the empirical results, section 6 describes other related work, section 7 contains a discussion of the results, and finally the conclusion and future work are presented in section 8.

2 Materials

One of the main goals was to test how good the current state of the GeneTUC parser is. To do this, some manually inspected parsed text is needed, and that is exactly what the new syntactically enhanced GENIA corpus is [19]. It consists of text from MEDLINE (see subsection 2.1), and provides a gold standard that can be used both for training and testing the GeneTUC application. See Subsection 2.2 for more details.

2.1   MEDLINE

Medline[1] is an online collection of more than 14 million abstracts by now (November 2005). The abstracts are collected from a set of different medical journals by the US National Institutes of Health (NIH). NIH grants academic licenses for PubMed/MEDLINE for free to anyone interested. When it was downloaded in September 2004, the academic package included a local copy of 6.7 million abstracts, out of the 12.6 million entries that were available on their web interface at that time.

2.2   GENIA Tree Bank (GTB)

It was decided to use the GENIA Tree-Bank (GTB) corpus[2] for training and testing of GeneTUC. GTB consists of 500 abstracts from the GENIA corpus which consists of 2000 abstracts from MEDLINE. These 500 abstracts have been parsed, manually inspected and corrected to ensure that they contain the expected parse result for every single sentence. The format of the annotation is a slightly modified Penn Tree Bank-II format. The GTB is split into GTB200 with 200 abstracts and GTB300 with 300 abstracts. We used GTB300 as a training set, and GTB200 as test set to calculate the precision and recall scores for parsing of *unseen* text. It should be pointed out that GTB is still in a beta-release state, which means that it still contains some errors, and some manual inspection of the results are needed to determine if this has a great influence on the measured numbers.

A list of all composite terms in the GTB was also used as input to the system. This was done to ensure that the parsing performance was measured without being influenced by bad tokenization, which is handled by another module, namely the lexical analysis module, in GeneTUC.

3   GeneTUC

GeneTUC is on the way to be a full-fledged Question Answering system, but the coverage is still low. Figure 1 shows the general information flow in the TUC systems. The input sentence can be either a fact for example from a Medline abstract or a question from the user. The analysis is the same in both cases, but the answer will have two different forms. In the case of a factual input sentence, the facts are coded in a first order event logic form called TUC Query Language (TQL) and then stored in the Knowledge Base (KB) of the system. This is shown in Example 1, above the line. Later, when someone inputs a question, the question will also first be coded using TQL, but either the subject or one of the objects in the sentence may then be wildcards that should be matched against facts in the existing KB.

---

[1] http://www.ncbi.nlm.nih.gov/entrez/query.fcgi
[2] http://www-tsujii.is.s.u-tokyo.ac.jp/GENIA/topics/Corpus/GTB.html

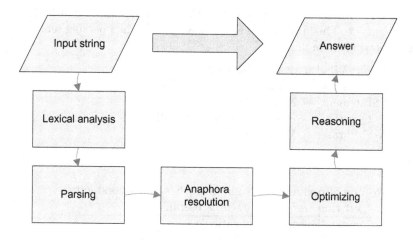

Fig. 1. Data flow in the TUC System

Example 1.
$$\frac{\begin{array}{c}\text{Statement : "CCK activates Gastrin."}\\ \text{Update to KB : activate(cck,gastrin).}\end{array}}{\begin{array}{c}\text{Question : "What activates Gastrin?"}\\ \text{Answer : "CCK"}\end{array}}$$

In this case it is very obvious that "What" is the placeholder for the answer, and also that it must be substituted with "CCK" to match the existing fact. So even if this is a very simple example, the method works in the same way also for more complex sentences. The only requirement is that the question is stated in a similar grammatical form as the factual statement.

3.1 GeneTUC Lexical Analysis

The lexical analysis in GeneTUC changes the input sentences from a long list of characters into tokens (words) and sentence delimiters. The current set of sentence delimiters includes the following:

Period	Colon	Semi Colon	Question Mark	Exclamation Mark
.	:	;	?	!

In the process of making the tokens, no distinction is made between upper and lower case characters, so the input to the syntactical analysis is a set of all lower case tokens.

3.2 GeneTUC Grammar and Syntactical Analysis

The GeneTUC grammar is what we call ConSensiCAL. That means it is a Context Sensitive Categorial Attribute Logic grammar formalism. It is based on the Prolog Definite Clause Grammar (DCG) with a few extensions to handle categorial movement and gaps etc. See [5] for more details on the Prolog programming language for Natural Language Processing (NLP).

Categorial Grammar. TUC is inspired by Categorial Grammar which allows *gaps* in the sentence. This mechanism is very easy to use when parsing sentences like in the following examples:

Example 2.

Input: What activates Gastrin?

Grammar for Question, using Statement:

    Statement → NounPhrase VerbPhrase

    Question → *what* Statement\NounPhrase

VerbPhrase → Verb NounPhrase

    . . .

Example 3.

Input:

Results of preliminary studies, which we have conducted, suggest that use of this agent is useful.

Grammar (Forward Application):

    NounPhrase → Det Nominal RelativeClause

    RelativeClause → RelativePron Statement/NounPhrase

RelativePronoun → *that*|*which*|*who*

    . . .

Example 4.

Input: A gene signal that results in production of proteins occurs.

Grammar (Backward Application):

    Statement → NounPhrase RelativeClause

RelativeClause → RelativePronoun Statement\NounPhrase

    . . .

Example 5.

Input: A gene signal resulting in protein production occurs.

Grammar for Gerund:

RelativeClause → Verb-*ing* RelativeClause*that*Verb-*s*

    . . .

Example 2 shows how the *What-Question* from Example 1 can be parsed using the existing grammar rules for Statement. It states that a *"what-question"* consists of the word "what" followed by a *Statement*, which is missing the leading *Noun Phrase (NP)*. This kind of Categorial *movement* makes it possible to connect the missing NP in the question ("what") with the actual NP in a corresponding fact statement ("CCK"), and then give a correct answer to the natural language query. This use of Backward (\) and Forward (/) application also reduces the number of grammar rules needed, since every new rule for statements implicitly creates corresponding new rules for questions.

In Example 3 the use of Forward application is shown. In GeneTUC, Forward application also includes Inward application, so "S/NP" also means that the NP can be missing anywhere in the Statement.

In Example 4, Backward application is used to define a Relative Clause. The missing NP in the Relative Clause can be found by going back to the NP that is preceding the Relative Clause.

Example 5 shows a different form of the sentence from Example 4. With the help of Backward application only one rule is needed to change this *gerund* sentence into a RelativeClause that can then be parsed by the grammar given in Example 3. This rule is context sensitive, meaning that *ing*-verbs like "resulting" can only be substituted by "that verb-*s*" phrases, like "that results", when the parser is already expecting to see a RelativeClause.

3.3   Reducing the Parsing Time

In GeneTUC parsing time is reduced by the use of *cuts* in the Prolog code. This means that once a specific rule, for example *Noun Phrase (NP)*, has been successfully applied to a part of the input sentence, this part of the sentence is *committed* and can not be parsed again even if the following rules causes the parser to fail. Usually, failing on one possible parsing attempt would cause the parser to back-track and use the rule on a different span of words to produce a different and successful NP. This kind of backtracking can be very computation-ally expensive, especially with highly ambiguous input, so *cuts* greatly reduces the parse time. The cuts are placed manually in strategic places in the code, based on experience from previous parsing of *run-away* sentences. Usually, the cuts do not affect the final result from the parser, but some phenomena can cause the parser to fail because the assumed partial parsing result before the cut is incompatible with the rest of the sentence. One such phenomenon, which is hard to parse even for humans, is *garden path sentences* [13].

4   Methods

The main goal of this research was to evaluate the GeneTUC parser on the GENIA corpus. Since GeneTUC and GENIA were not made using any com-mon grammar standard, a lot of modifications in GeneTUC were needed. These adaptations can be thought of as a (manual, not statistical) training process for GeneTUC, but in order to measure how the GeneTUC parser will perform on unseen data, different parts of the GENIA Tree Bank (GTB) was used for training and testing, i.e. we used 300 abstracts (GTB300) for training and the remaining 200 abstracts (GTB200) for testing.

The training phase of the project is described in the following subsections, and includes the following tasks:

- Dictionary building. Adding all terms from GENIA to the GeneTUC dictio-nary.
- Ontology building. Mapping from GENIA to GeneTUC dictionary classes.
- Adding other missing words manually, with the help of Bioogle.
- Input new verb templates, based on predicate argument structures seen in GENIA.

4.1   Updating GeneTUC Lexicon from GENIA

Since the goal is to test the parser, errors connected to the Tokenizer, POS tagger or other parts of the system should be removed. The ideal approach would be to use the tokenized and POS-tagged version of GENIA as input to GeneTUC, but this was not feasible since GeneTUC is based on plain ASCII-text input. Also, it would take more manpower/time than available in this project to split the tight connection between tokenizing, tagging and parsing in TUC, just to test if it would be useful to do so later. Instead, the GENIA multi-word-terms were added to the GeneTUC dictionary, trying to guide it into using the same tokenization as in the GENIA gold file. This was only successful in around 20% of the sentences, so we reduced the test set to only include sentence that were similarly tokenized and tagged by GENIA and GeneTUC.

During the process of importing all the terms from GENIA into the TUC, several considerations had to be made:

1. *Plural Forms.* Plural words were changed into their singular (stem) form by simple rules like: remove the final s from all words. Exceptions to this simple rule had to be made for words like virus (already singular), viruses (remove -es) and bodies (change -ies to y).
2. *Proper Names or Common Nouns.* Another point is that plural forms should exist only as ako[3] relations (class concepts), and *not* as isa[4] relations (proper names).
3. *Duplicate Entries.* Changing plural forms into singular forms often leads to duplicate entries in the dictionary, since the singular form
4. *Short Ambiguous Terms.* The title sentence "Cloning of ARE-containing genes by AU-motif-directed display" causes a problem since TUC does not distinguish "ARE" and "are". Words like "are", "is", "a" and so on are therefore removed from the dictionary.

4.2   Updating the GeneTUC Semantic Network

As mentioned in the introduction, GeneTUC is a deep parser, requiring that all the input words are already in the dictionary. In order to compare just the parsing performance of GeneTUC with other systems, other error sources such as incomplete lexical tagging was reduced by importing all named entities from GENIA to GeneTUC. When new words are added to GeneTUC, it is also necessary to specify which semantic class they belong to, so a mapping between GENIA ontology and the ontology of GeneTUC was needed (Figure 2). One alternative way was to simply add all the ontology terms of GENIA (37) to GeneTUC, but many of the terms were already present in both systems, with slightly different classifications. We could also have changed the GeneTUC ontology terms to match those of GENIA, but that would have made many of the existing verb templates in GeneTUC useless or wrong, making this approach unappealing. The final choice was to create a mapping from GENIA ontology terms to existing GeneTUC ontology terms, as shown in Figure GENIA ontology. The GENIA term "other_name" and the corresponding GeneTUC term "stuff" are "bag" definitions, meaning that no effort was made to distinguish the terms that

---

[3] ako = A Kind Of (subclass of a class).

[4] isa = Is A (instance of a class).

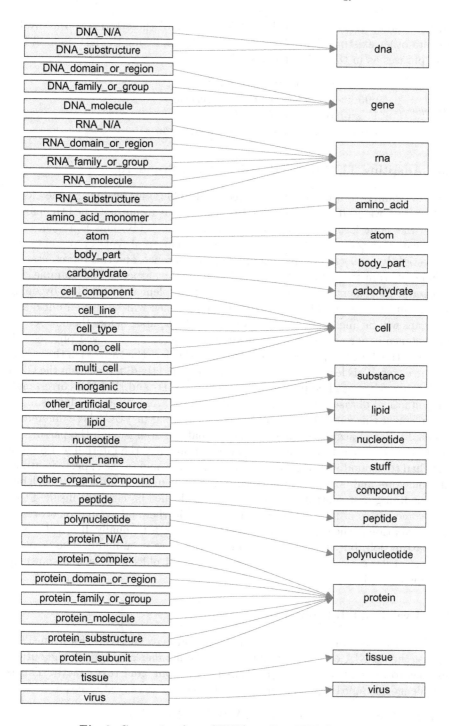

Fig. 2. Conversion from GENIA to GeneTUC Ontology

did not belong to one of the other classes. Many of these terms can easily be put into other existing GeneTUC classes, just by matching the last noun in the noun phrases as in the following example:

Example 6.

"nf_kappa_b_activation" *ako* activation
'0_95_kb_id_3_transcript' *ako* transcript
'17_amino_acid_epitope' *ako* epitope
asp_to_asn_substitution *ako* substitution

4.3   Adapting TUC to GTB/PTB Syntax Standard

Since we wanted to use the GENIA Tree Bank (GTB) for evaluating the Gene-TUC parser, we needed to make sure that the output from the GeneTUC parser was in the same format as the parse trees from the GTB. This is a somewhat complicated process, since the TUC parser uses an internal syntax representation that is tightly connected to the semantics of the sentence, and this representation is different from the GTB syntax in a few important aspects. For example, the Categorial movement and gap mechanisms are implemented in TUC by doing parsing and reparsing. That means that the sentence will be parsed once, and then gaps will be filled with the syntax from the first parse, before the new resulting sentence is parsed again. This means that traces of the moved phrases will appear both where the phrase was originally, and where the gap was in the resulting parse tree. This leads to parse trees that look slightly different from the GTB parse trees, where each gap is given an Identifier (ID) and then the corresponding syntactical phrase is given the same ID-number. As long as no effort is made to implement this gap-ID system of the GTB grammar in TUC, these differences will lead to lower accuracy values in the evaluation, even though the parsing result is actually correct. To prevent this from happening, the internal workings of TUC had to be modified to produce output exactly equal to the expected output, and some pre- and post-processing scripts had to be made in order to smooth out the remaining systematical differences that could not be changed inside TUC. Still, some traces of these problems may be left in the final evaluation scores.

The creation of the grammar is currently done 100% manually. It is a slow and long-lasting job, but it ensures that all the rules are meaningful. The creation of a new rule is always rooted in the existence of old rules, as was shown in Examples 2 & 3.

4.4   EvalB and Tokenization

EvalB[5] was used for calculation of precision and recall scores for GeneTUC against the GENIA corpus. It requires that the number of tokens in the output text has to match the number of tokens in the input/gold text exactly. This is a challenge to GeneTUC, since it ignores the characters listed in Example 7.

Example 7. Ignored Characters: " : , & % { } ⟨ ⟩ [ ] (...)

---

[5] http://nlp.cs.nyu.edu/evalb/

Also, single tokens (like IL-2) are sometimes turned into two or three separate tokens ("il", "-" and "2"), because of hyphens. This happens when the word is not specifically defined in the dictionary as being just one word/token. Since the GTB is already tokenized and stored in XML format, the correct tokenization is known. The challenge is to ensure that TUC produces the expected output, even if the internal modules are using different tokens. Other features of GeneTUC that makes the comparison difficult is that some *noisewords* are removed from the text, and long Noun Phrases can sometimes be substituted with Canonical Identifiers.

There are two obvious solutions to this problem: The first is to use the tokenized version of GTB, instead of the plaintext version. The problem then, is that we have to circumvent the tokenization module (and lexical analysis) in TUC, and this might introduce problems in the later modules, for example because of ignored characters that were previously handled by the lexical module. Another example of problems introduced if the original tokenization is used, is parentheses with their contents. In the current implementation all level-1 parentheses are removed together with everything inside them, since this is usually ungrammatical constructions.

The second solution to the tokenization problem is to make a new plaintext version of the text, from the tokenized xml-version. In the new plain text version, all tokens containing hyphens and other troublesome characters, will be substituted by a new token using underscore (_) or some other character instead of hyphen, so that the lexical module does not split these token into extra tokens. In the opposite case, where the gold text contains more tokens than what TUC produces, we have to introduce some dummy tokens. These tokens can then act as placeholders for tokens that TUC ignores (and removes), like parentheses with all their content/tokens.

4.5   EvalB Comparing Syntax Trees

Using the tokens in the sentence as basis for scoring, EvalB performs a strict evaluation. Any case of tokenization different from the "golden" tokenization renders the parse incomparable; the sentences where the number of golden tokens and test tokens are unequal lead to an *error*. The same is true for those where the golden token and test token are character wise different. If the number of tokens equals zero (i.e. the sentence did not parse successfully), the sentence is *skipped*. Both *skip-* and *error*-sentences are ignored when calculating the score. The program provides a tolerance limit for how many incomparable sentences that are ignored.

Bracketing is measured from token[m] to token[n], where a *match* means those brackets covering the correct tokens, and having a correct *label*. The matches enable measurements of:

- Recall (the ratio between matched brackets and all brackets in the gold file)
- Precision (the ratio between matching brackets and all brackets in the test file)

- Crossing (the number of test-file-brackets exceeding the span of a matching bracket in the gold file, divided by the total number of brackets in test file)
- Tagging accuracy (the ratio of correct labeled tokens over the total number of tokens)

EvalB performs strict evaluation of the parse, as it originally was intended as a solemn bracketing evaluation program. Bracketing scores of GeneTUC may be reduced because of a right-orientation implied by the grammar of TUC (always preferring right-attachment unless it is semantically erroneous).

5   Results

This section shows the results from the training and test phases. Table 1 shows how much the performance of GeneTUC increased when the dictionary was extended with all the terms from GENIA. Table 2 shows that there was no significant difference in parsing scores between importing all the GENIA terms (36.692) or just the terms from GTB200 that were reported as unknown by GeneTUC

Table 1. Statistics parsing attempts before and after adding GENIA dictionary

Measurement	Dictionary	
Description	Original	+GENIA
Number of sentences:	2591	2591
Successful parses:	318	1531
Success rate:	12.3%	59.1%
Sources of Failure		
Dictionary:	1989	68
Grammar:	520	1126
Time out:	32	144
Processing time:	0.5 hrs	4.5 hrs

Table 2. Test results from EvalB

Measurement	Dictionary	
Description	+GENIA	+GTB200
Number of sentences	1759	1759
Error sentences	518	565
Skip sentences	1037	843
Valid sentences	204	351
Bracketing Recall	49.8%	53.9%
Bracketing Precision	69.0%	70.6%
Complete match	0.49%	1.14%
Average crossing	1.27	1.47
No crossing	47.1%	44.7%
2 or less crossing	79.9%	79.5%
Tagging accuracy	82.0%	86.0%

(8.175). In terms of input to EvalB, it was possible to compare almost twice as many sentences when only the GTB200 dictionary was used. This is mainly because GeneTUC was given fewer chances to rewrite complex multi-word tokens, and thereby creating better accordance between the output and the gold file.

6 Related Work

Other recent and related IE techniques for biomedical literature includes systems using dynamic programming [8] or supervised machine learning [22] to find protein-protein-relations in molecular biology texts. The machine learning approach uses both parse trees and dependent tree structures as features, but they report that simple lexical features contribute more to the promising F-measure of 90.9. Other systems use predicate-argument structure patterns [23] or new self made architecture [21] to do more general Information Extraction from this kind of free text sources.

6.1 GeneTUC, Bioogle and GProt

This paper showed how important a proper dictionary is to this kind of semantic parsers. A new way to automatically build dictionaries with ontology information is presented as Bioogle in [18]. Bioogle[6] is a simple system that uses Google to determine the semantic class of a word, for example "CCK is a protein", so that it can be added to the semantic hierarchy (or dictionary) in a correct way.

GProt [17], is another application that is built on top of the Google API, like Bioogle. GProt[7] provides a way of automatically extracting information from the (biomedical research) literature. Most of the literature is already indexed in MEDLINE, and therefore also by Google and other major search engines. See [16] for more details and a description of how to access the online versions of Bioogle and GProt.

7 Discussion

This section points out some of the lessons learned during the parsing project. This includes remarks about titles as a different sentence type and a discussion about the results presented in the two previous sections.

7.1 Sentence Types

MEDLINE (GTB) contains two fundamentally different sentence types: Titles and normal sentences. The titles are special, because they sometimes just state the object of the experiment, without the subject and verb phrase that should have started the sentence. Subject and verb-less sentences were already handled by GeneTUC before, but during this work we added a special "\title"-tag for

---

[6] http://furu.idi.ntnu.no:8080/bioogle/
[7] http://furu.idi.ntnu.no:8080/gprot/

the titles, so that we can implement some special processing of titles later. The first function we added to the "\title"-tag was resetting the temporary anaphoric database, so that terms like "the protein", "this" and "which" do not map to names or events in any previously parsed abstracts.

7.2 Comparing Different Systems

It turned out that evaluating the GeneTUC parser on a PTB gold standard file was harder than first expected. The main reason for this is that TUC was never meant to output PTB style tags in the first place. Also, there is not always a clear boundary between lexical, syntactical and semantic analysis. Of course, there are both advantages and disadvantages to this approach.

The problem that we encountered because of the tight connection between the modules in GeneTUC, is that it is very hard to construct output with the exact number of tokens as in the input text. TUC is based on receiving plain text input, and does its own tokenization and optimization of the text before passing it on to the parser. We could perhaps have used the already tokenized text as input, but this would introduce the parser to problems it is not meant to handle in the current configuration of the system. It would be much easier to cooperate with other researchers if the modules of GeneTUC were truly separate from each other, but it can also be argued that the good performance by a text processing system like this is really dependent on tight communication between the modules.

Tokenization is usually done before, and separate from, parsing, but sometimes it is necessary to do preliminary parsing in order to determine word and sentence boundaries. Parsing is usually done before semantic analysis, but sometimes it is important to know the semantic properties of a word in order to reduce the ambiguity, and thereby the parsing time. Maybe the time has come to start integrating the different modules more? This will require some effort to ensure that cooperation between different researchers is still possible, for example through the use of new standards/protocols for future parsers.

8 Conclusion

There is a great need for systems that can support biologists (and any other research) in dealing with the ever increasing information overload in their field. This project has proven that both the Google API and the GeneTUC systems are important pieces that can play a role in making the dream of real automatic Information Extraction come true in the not so distant future.

The precision and recall scores achieved by GeneTUC on general parsing are not very high compared to pure parsers like [4,10,7], but that does not mean that these systems are better than GeneTUC, because GeneTUC also performs deeper analysis such as anaphora resolution [2,14]. The other systems consist of Context-Free Grammar (CFG) parsers that give only phrase structures as output. There are also some systems that use CCG parsers [3] or HPSG parsers [11] that can give predicate argument structures in addition to phrase structures, but they still do not perform anaphora resolution or question-answering, like GeneTUC does.

Acknowledgements

The first author would like to thank all the people who made the writing of this chapter possible. Especially, Professor Tsujii who invited me to his lab in Tokyo, and all his brilliant co-workers who helped me with anything related to the GENIA corpus. Much gratitude also goes to my co-supervisors Amund Tveit and Astrid Lægreid for continuous support, and for pointing out good opportunities.

References

1. Razvan Bunescu, Ruifang Ge, Rohit J. Kate, Edward M. Marcotte, Raymond J. Mooney, Arun Kumar Ramani, and Yuk Wah Wong. Comparative Experiments on Learning Information Extractors for Proteins and their Interactions. *Journal Artificial Intelligence in Medicine: Special Issue on Summarization and Information Extraction from Medical Documents*, 2004.
2. J. Castano, J. Zhang, and J. Pustejovsky. Anaphora resolution in biomedical literature. In *International Symposium on Reference Resolution*, 2002.
3. Stephen Clark, Julia Hockenmaier, and Mark Steedman. Building Deep Dependency Structures with a Wide-Coverage CCG Parser. In *Proceedings of ACL'02*, pages 327–334, 2002.
4. Andrew B. Clegg and Adrian J. Shepherd. Evaluating and integrating treebank parsers on a biomedical corpus. In *Proceedings of the ACL Workshop on Software 2005*, 2005.
5. Michael A. Covington. *Natural Language Processing for Prolog Programmers*. Prentice-Hall, Englewood Cliffs, New Jersey, 1994.
6. J. Cowie and W. Lehnert. Information Extraction. *Communications of the ACM*, 39(1):80–91, January 1996.
7. Tadayoshi Hara, Yusuke Miyao, and Jun'ichi Tsujii. Adapting a probabilistic disambiguation model of an HPSG parser to a new domain. In *IJCNLP 2005: Second International Joint Conference on Natural Language Processing*, 2005.
8. Minlie Huang, Xiaoyan Zhu, Yu Hao, Donald G. Payan, Kunbin Qu, and Ming Li. Discovering patterns to extract protein-protein interactions from full texts. *Bioinformatics*, 20(18):3604–3612, Dec 12 2004.
9. Tor-Kristian Jenssen, Astrid Lægreid, Jan Komorowski, and Eivind Hovig. A literature network of human genes for high-throughput analysis of gene expression. *Nature Genetics*, 28(1):21–28, May 2001.
10. Matthew Lease and Eugene Charniak. Parsing biomedical literature. In *Second International Joint Conference on Natural Language Processing (IJCNLP'05)*, 2005.
11. Yusuke Miyao and Jun'ichi Tsujii. Deep linguistic analysis for the accurate identification of predicate-argument relations. In *Proceedings of COLING 2004*, pages 1392–1397, 2004.
12. R. O'Donovan, M. Burkea, A. Cahill, J. van Genabith, and A. Way. Large-Scale Induction and Evaluation of Lexical Resources from the Penn-II Treebank. In *Proceedings of the 42nd Annual Meeting of the ACL.)*, pages 368–375, Barcelona, Spain, July 21-26 2004. Association for Computational Linguistics.

13. Lee Osterhout, Phillip J. Holcomb, and David A. Swinney. Brain potentials elicited by garden-path sentences: Evidence of the application of verb information during parsing. *Journal of Experimental Psychology: Learning, Memory, and Cognition*, 20(4):786–803, 1994.

14. J. Pustejovsky, J. Casta, J. Zhang, B. Cochran, and M. Kotecki. Robust relational parsing over biomedical literature: Extracting inhibit relations. In *Pacific Symposium on Biocomputing*, 2002.

15. Rune Sætre. Natural Language Processing of Gene Information. Master's thesis, Norwegian University of Science and Technology, Norway and CIS/LMU München, Germany, April 2003.

16. Rune Sætre, Martin T. Ranang, Tonje S. Steigedal, Kamilla Stunes, Kristine Misund, Liv Thommesen, and Astrid Lægreid. Webprot: Online Mining and Annotation of Biomedical Literature using Google. In Tuan D. Pham, Hong Yan, and Denis I. Crane, editors, *Advanced Computational Methods for Biocomputing and Bioimaging*. Nova Science Publishers, New York, USA, 2006.

17. Rune Sætre, Amund Tveit, Martin Thorsen Ranang, Tonje Strømmen Steigedal, Liv Thommesen, Kamilla Stunes, and Astrid Lægreid. GProt: Annotating Protein Interactions Using Google and Gene Ontology. In *Lecture Notes in Computer Science: Proceedings of the Knowledge Based Intelligent Information and Engineering Systems (KES2005)*, volume 3683, pages 1195 – 1203, Melbourne, Australia, August 2005. KES 2005, Springer.

18. Rune Sætre, Amund Tveit, Tonje Strømmen Steigedal, and Astrid Lægreid. Semantic Annotation of Biomedical Literature using Google. In Dr. Osvaldo Gervasi, Dr. Marina Gavrilova, Dr. Youngsong Mun, Dr. David Taniar, Dr. Kenneth Tan, and Dr. Vipin Kumar, editors, *Proceedings of the International Workshop on Data Mining and Bioinformatics (DMBIO 2005)*, volume 3482 (Part III) of *Lecture Notes in Computer Science (LNCS)*, pages 327–337, Singapore, May 9-12 2005. Springer-Verlag Heidelberg.

19. Yuka Tateishi, Akane Yakushiji, Tomoko Ohta, and Jun'ichi Tsujii. Syntax Annotation for the GENIA corpus. In *Proceedings of the IJCNLP 2005*, Korea, October 2005.

20. Amund Tveit, Rune Sætre, Tonje S. Steigedal, and Astrid Lægreid. ProtChew: Automatic Extraction of Protein Names from . In *Proceedings of the International Workshop on Biomedical Data Engineering (BMDE 2005, in conjunction with ICDE 2005)*, pages 1161–1161, Tokyo, Japan, April 2005. IEEE Press (Electronic Publication).

21. Aditya Vailaya, Peter Bluvas, Robert Kincaid, Allan Kuchinsky, Michael Creech, and Annette Adler. An architecture for biological information extraction and representation. *Bioinformatics*, 21(4):430–438, 2005.

22. Juan Xiao, Jian Su, and GuoDong Zhouand ChewLim Tan. Protein-protein interaction extraction: A supervised learning approach. In *Semantic Mining in Biomedicine (SMBM)*, 2005.

23. Akane Yakushiji, Yusuke Miyao, Yuka Tateishi, and Junichi Tsujii. Biomedical information extraction with predicate-argument structure patterns. In *Semantic Mining in Biomedicine (SMBM)*, 2005.

24. Hong Yu, Vasileios Hatzivassiloglou, Carol Friedman, Andrey Rzhetsky, and W. John Wilbur. Automatic Extraction of Gene and Protein Synonyms from MEDLINE and Journal Articles. In *Proceedings of the AMIA Symposium 2002*, pages 919–923, 2002.

Gene Relation Finding Through Mining Microarray Data and Literature

Hei-Chia Wang*, Yi-Shiun Lee, and Tian-Hsiang Huang

Institute of Information Management, National Cheng Kung University,
Tainan, 701, Taiwan
hcwang@mail.ncku.edu.tw, rog@iis.sinica.edu.tw, huangtx@gmail.com

Abstract. Finding gene relations has become important in research, since finding relations could assist biologists in finding a genes functionality. This article describes our proposal to combine microarray data and literature to find the relations among genes. The proposed method tries emphasizes the combined use of microarray data and literature rather than microarray data alone. Currently, many scholars use clustering algorithms to analyze microarray data, but these algorithms can find only the same expression mode, not the transcriptional relation between genes. Moreover, most traditional approaches involve all-against-all comparisons that are time-consuming. To reduce the comparison time and to find more relations in a microarray, we propose a method to expand microarray data and use association-rule algorithms to find all possible rules first. With its literature text mining, our method can be used to select the most suitable rules. Under such circumstances, the suitable gene group is selected and the gene comparison frequency is reduced sharply. Finally, we can then apply dynamic Bayesian network (DBN) to find the genes interaction. Unlike other techniques, this method not only reduces the comparison complexity but also reveals more mutual interactions among genes.

1 Introduction

The Human Genome Project (HGP) has been in operation for a decade. The project was originally planned to be completed in 15 years, but rapid technological advances accelerated its completion in 2003. The HGP's ultimate goal was to identify all the functions of the 30,000 human genes and render them accessible for further biological study. After the announcement that all human genes had been sequenced, the next stage was to investigate functional genomics [7,5]. To achieve this purpose requires the application of a powerful technique. Microarray technology is one of them.

The development of microarray technology has been phenomenal in the past few years. It has become an important tool in many genomics research laboratories because it has revolutionized the approach to finding the relations among genes. Instead of working on a gene-by-gene basis, scientists can use microarray to study tens of thousands of genes at once. Many benefits result from this technology. One is determining how the expression of a particular gene might affect

C. Priami et al. (Eds.): Trans. on Comput. Syst. Biol. V, LNBI 4070, pp. 83–96, 2006.
© Springer-Verlag Berlin Heidelberg 2006

the expression of other genes. Another is determining the gene network or what genes are expressed as a result of certain cellular conditions, for example, genes expressed in diseased cells that are not expressed in healthy cells.

Currently, the most widely used algorithm for analyzing microarray data is hierarchical clustering [4,20,18]. Although this method has been successful in showing that genes participating in the same biological processes have the same expression profiles, several problems remain. First, genes that are biologically related often are not related in their expression profiles and hence will not be clustered together [21,2]. Second, clustering genes into disjoint clusters will not capture the gene products that participate in more than one biological process [1,2]. Third, relationships among the different members of a group cannot be inferred [1]. Finally, the all-against-all comparison in clustering is time-consuming.

Here, we describe our attempt to solve the above problems with an association-rule algorithm to filter possible related genes first, which can reduce the number of candidate genes. In addition to microarray data, literature is also widely used to find gene relations [16,19]. To take both resources into consideration, we propose a method to combine literature and microarray data to construct a gene network. When the candidate genes have been selected, we apply a dynamic Bayesian network (DBN) to find the genes interaction. Since using a DBN for all genes is time-consuming, irrelevant genes should be filtered to reduce the impact [23]. In this research, we use an association-rule algorithm in the first step to filter irrelevant genes to reduce the impact and find extra information. In this stage, since the microarray data we used was limited and the association-rule algorithm requires a lot of data, we thought that variation over time may reveal some association of genes. Therefore, we designed an across time-finding algorithm to expand the microarray data. When the association rules were found, genes could be listed under more than one rule. Thus, a gene may be involved in more than one syn-expression group. In this stage, determining the most suitable rule was a challenge. In the proposed method, literature is then used to assist the search for the most suitable rule. Finally, DBN uses candidate genes from the recommend rules for constructing a network.

2   Background

The model uses 3 important concepts,: association rule by an association-rule algorithm, text mining by vector space model (VSM), and DBN. We explain them briefly in the following sections.

2.1   Association Rule

An association rule is presented in the form $LHS \Longrightarrow RHS$, where LHS and RHS both are "itemsets". Itemsets can be defined in terms of "transactions", which in the retail industry refers to customer transactions (a customer purchases one or more items at the checkout counter in a single transaction). Here we adopt the definition for itemset and association rules from [3] to implement our program:

Definition 1

1. Given a set S of items, any nonempty subset of S is called an itemset.
2. Given an itemset I and a set T of transactions, the "support" of I with respect to T, denoted by support$T(I)$, is the number of transactions in T that contain all the items in I.
3. Given an itemset I, a set T of transactions and a positive integer α. I is a "frequent itemset" with respect to T and α if support$T(I) > \alpha$. We refer to α as the "minimum support".

Definition 2

1. An "association rule" is a pair of disjointed itemsets. If LHS and RHS denote the two disjoint itemsets, the association rule is written as $LHS \Longrightarrow RHS$.
2. The "support" of the association rule $LHS \Longrightarrow RHS$ with respect to a transaction set T is the support of the itemset $LHS \cup RHS$ with respect to T.
3. The "confidence" of the rule $LHS \Longrightarrow RHS$ with respect to a transaction set T is the ratio of support $(LHS \cup RHS)$ to support (LHS).

2.2   Text Mining for Finding Gene Relations

This proposed method tries to not only find the gene relation from microarray data but also mine the literature. The text mining is based on the information retrieval to find the term weight and term correlation in each piece of literature. In the proposed method, the term in the literature will be weighted, and the VSM is used to find the similarity of term vectors.

The VSM was first proposed by [15]. It assigns each term a weight and presents a document in a vector $d_j = [w_{1,d_j}, w_{2,d_j}, w_{3,d_j}, \cdots, w_{t,d_j}]^T$. Each query is thought of as a vector as well. Through set the query term to the other vector $q = [w_{1,Q}, w_{2,Q}, w_{3,Q}, \cdots, w_{n,Q}]^T$ to find the similarity between the documents and the query. The similarity may apply a cosine similarity or inner product. In this implementation, cosine similarity is used. The similarity function is shown as equation (1).

$$sim(q, d_j) = \frac{d_j \cdot q}{\|d_j\| \times \|q\|} = \frac{\sum_{i=1}^{t} w_{i,d_j} \times w_{i,q}}{\sqrt{\sum_{i=1}^{t} w_{i,d_j}^2} \times \sqrt{w_{i,q}^2}} \tag{1}$$

The other issue in the VSM is how to find each terms weight in a document and query. The popular way uses tf (term frequence) or tf-idf (term frequency and inverse document frequency). In this application, tf-idf is applied for calculating the weight of each term in a document. The equation is shown in equation (2), where $w_{i,d}$ is term weight and $freq_{i,d}$ is the frequency of the term in a document. N is the total document number and n_i is how many documents contain the term.

	w_1	w_2	w_3	w_4	w_5	w_6	$\cdots\cdots\cdots$	w_{t-2}	w_{t-1}	w_t
q	1	0	0	3	0	1	$\cdots\cdots\cdots$	0	1	0

d_j	0	1	0	2	0	2	$\cdots\cdots\cdots$	1	1	0

$$w_{i,d_j} = \frac{freq_{i,d_j}}{\max\limits_{l} freq_{l,d_j}} \times \log \frac{N}{n_i} \qquad (2)$$

2.3 Dynamic Bayesian Network

DBNs can be viewed as an extension of Bayesian networks (BNs). In contrast to BNs, which are based on static data, DBNs use time series data for constructing causal relationships among random variables. Suppose that we have n microarrays and each microarray measures expression levels of p genes. The microarray data can present $n \times p$ matrix. $X = (x_1, x_2, \cdots, x_n)^T$ whose ith row vector $x_i = (x_{i1}, x_{i2}, \cdots, x_{ip})^T$ corresponds to a gene expression level vector measured at time t. The DBN model assumes that each genes responses depend only on the status of the previous time unit. In other words, the status vector of time i (in Figure 1) depends only on that of time i-1 [10,11,12].

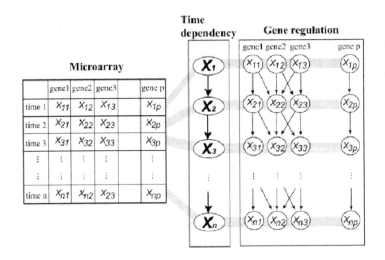

Fig. 1. DBN model

On the basis of this assumption, the joint probability can be decomposed as equation (3):

$$P(X_{11}, \cdots, X_{np}) = P(X_1)P(X_2|X_1) \times \cdots \times P(X_n|X_{n-1}) \qquad (3)$$

where $X_i = (x_{i1}, \cdots, x_{ip})^T$ is a p-dimensional random variable.

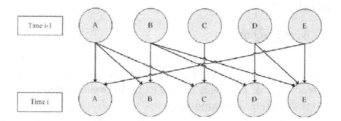

Fig. 2. DBN interaction diagram

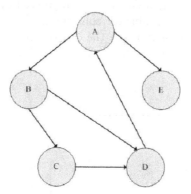

Fig. 3. DBN used to solve cycles

According to the time dependency, only forward edges (i.e., edges from time i-1 to i) are allowed in the network, seen in Figure 2.

With DBN, the interaction between genes can be divided into 2 slices (shown as Figure 2). Hence, use of DBN can solve the cycle situation that BN cannot [10,11]. From the structure in Figure 3, $P(X_i|X_{i-1})$ can be decomposed to equation (4).

$$P(X_i|X_{i-1}) = P(X_{i1}|P_{i-1,1}) \times \cdots \times P(X_{ip}|P_{i-1,p}) \tag{4}$$

where $P_{i-1,j} = (P_{i-1,1}^{(j)}, \cdots, P_{i-1,q_j}^{(j)})^T$ is a random variable vector of parent genes of jth gene at time $i-1$. We can then combine equations (3) and (4) for the DBN model as equation (5).

$$f(x_{11}, \cdots, x_{np}) = \prod_{i=1}^{n} \prod_{j=1}^{p} g_j(x_{ij}|p_{i-1,j}) \tag{5}$$

where $p_{oj} = \phi$.

3   Methodology

3.1   System Architecture

In this research, the expanded microarray data is considered as input data. The output information is the gene relations represented in a network chart. By these relations, the reactions of catalysis or restraint between genes can be obtained to help biologists find relations among genes.

The system contains a mining engine module, retrieval module, and a constructing network module as shown in Figure 4. The mining engine module uses an association rule to process gene data in a microarray to determine gene relations. Once the rules are generated, they are reordered by a retrieval module to find which rule is the most suitable for constructing a network module. Finally, the selected rules are used as input in the construction network module, which uses DBN to construct a gene network. The following sections explain these 3 modules in detail.

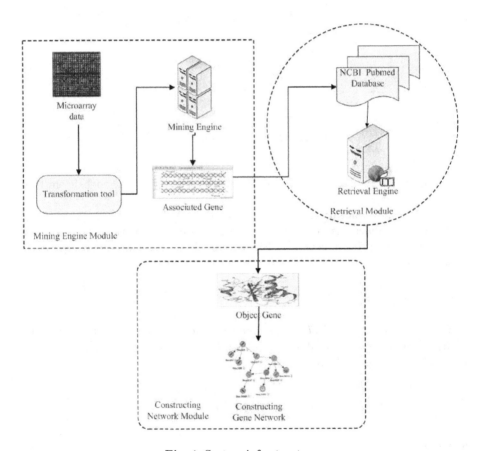

Fig. 4. System infrastructure

Table 1. Microarray data

	YBR160W	YER111C	YAL040C	⋯	YLR182W
$1Cy5$	294	541	1680	⋯	302
$1Cy3$	166	511	701	⋯	395
$2Cy5$	264	35	1032	⋯	275
$2Cy3$	213.5	377	523	⋯	370
⋯	⋯	⋯	⋯	⋯	⋯
$24Cy5$	25	53	20	⋯	24
$24Cy3$	17	48	15	⋯	22

Table 2. Microarray transporting data array

	YBR160W	YER111C	YAL040C	⋯	YLR182W
1-2	D	D	D	⋯	D
1-3	D	D	D	⋯	N
1-4	D	D	D	⋯	D
⋯	⋯	⋯	⋯	⋯	⋯
2-3	D	D	D	⋯	D
2-4	D	D	D	⋯	D
⋯	⋯	⋯	⋯	⋯	⋯
23-24	U	D	N	⋯	N

3.2 The Mining Engine Module

Currently, most microarray data are analyzed by a clustering algorithm, which can find only the same model and may cause some information loss (i.e., transcription factor loss). To find this information, we adopt time series microarray data to make the variation between 2 slices more meaningful. We first expand the microarray data to find variation between 2 slices, then we use an association algorithm to analyze the data to find the complete relations among genes. Table 1 shows a sample microarray data.

We then use the following steps to find rules by different time slices.

1. Produce all possible situations from different time slices.
2. 2. Using the data in Table 1, obtain the cy5 front value and background value for equation (6) to calculate the expression value to find the trend between the time slices. The results are shown in Table 2.
3. Use the results to run an association rule algorithm to determine possible associations.

$$Expression\_Value = log_2 \left[\frac{(Cy5_i(F) - Cy5_i(B)) - (Cy5_j(F) - Cy5_j(B))}{(Cy3_i(F) - Cy3_i(B)) - (Cy3_j(F) - Cy3_j(B))} \right]$$
(6)

After step 2, 3 possible levels could be found: "U" for "expression", "N" for "no reaction" and "D" for "deterrence". They are judged by the rule:

$Value = \log(Cy5/Cy3)$. When the expression value is **larger than 2**, the data will be transferred to "U" as the expression status. When the value is **between 1 and 2**, the data will be transferred to "N", which means that the gene has no significant reaction in this process. When the value is **less than 1**, the data will be transferred to "D", which means that the gene is in deterrence status at this time slice.

After this step, the association rule algorithm is applied to find the possible rules. In this example, we find that:

If YBR160W = D & YER111C = D &$\cdots$& YFL009W = D,
then YLR182W = D.

3.3   The Retrieval Module

This module uses literature from the National Center for Biotechnology Information (NCBI) PubMed to find gene relations. The module workflow is shown in Figure 5. Use of literature reconfirms which rules are reliable. This module is used to find suitable rules because many nonqualified rules will be found from association rules [9]. To find the high-quality association rule generated from the last module, relation-finding by literature search is used to confirm the suitability.

In the retrieval model, the TF-IDF model is used to reconfirm gene relations by finding their co-occurrence frequency in each article. This method is widely

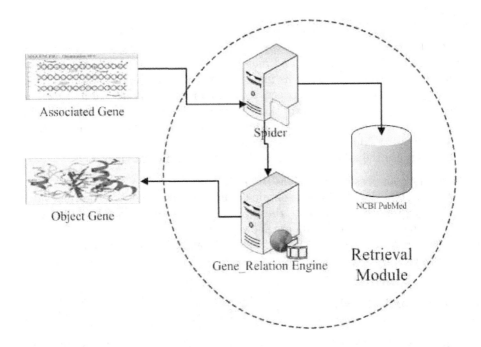

Fig. 5. Retrieval module

Table 3. Gene relation strength found from literatures

	YBR160W	YER111C	YAL040C	···	YLR182W
YBR160W	1	324	120	···	40
YER111C	324	1	100	···	50
YAL040C	120	100	1	···	225
···	···	···	···	···	···
YLR182W	40	50	225	···	1

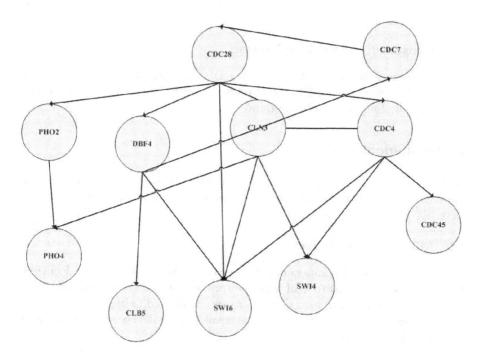

Fig. 6. A sample gene network

applied in research [17,8,13]. The method we used is modified from [17] and [13] to find gene relations by the equation (7).

$$Association[k][l] = \sum_{i=1}^{N} W_i[k] \times W_i[l] \quad k, l = 1, \cdots, m \tag{7}$$

where

$Association[k][l]$ are gene-relation strength, and $W_i[k]$ is gene k weight in article i.

To ignore the effect of total gene number, the weight of each rule is divided by the square of the gene number (equation 8)

$$Weight_k = \frac{\sum_{i=1}^{n} \sum_{j=1}^{n} Association[i][j]}{n^2}. \tag{8}$$

Continuing the rule found from the example, we can obtain the gene weights in Table 3.

We can then normalize the table value between 0 and 1 by equation 9.

$$Weight = \frac{x - \mu}{\sigma} \tag{9}$$

3.4   Constructing the Gene Network

After obtaining the most reliable rule, a DBN is then used to build a gene network. We select all genes in the rules and ignore the rest. Figure 6 shows a sample result from our evaluation data.

4   Implementation

Figure 7 is the deployment diagram of this system. It contains a mining engine server and information retrieval server for prepress and a constructing network server for finding the gene relations.

4.1   Mining Engine Server

This server contains 2 components, $Time\_Series$ and $Magnus\_Opus\_Version$. $Time\_Series$ is designed to deal with the time serious data of microarray. After expanding the data, $Magnum\_Opus\_Version$ [22] is used to find the association rule of these genes and an information retrieval server will check to find the most suitable rules for building a network.

4.2   Information Retrieval Server

Three components are used in the information retrieval server: $Spider$, $Gene\_To\_Gene\_Relation$, and $Integration\ Machine$. Spider is used to download all gene-related documents from the NCBI PubMed. $Gene\_To\_Gene\_Relation$ is used to find the gene relation strength. Finally, $Integrating\_Machine$ is used to find the most recommended rule.

4.3   Constructing Network Server

After obtaining the most recommended rule, BNT_DBMCMC [14] is applied for constructing a network server. The gene network for each rule is presented at this stage.

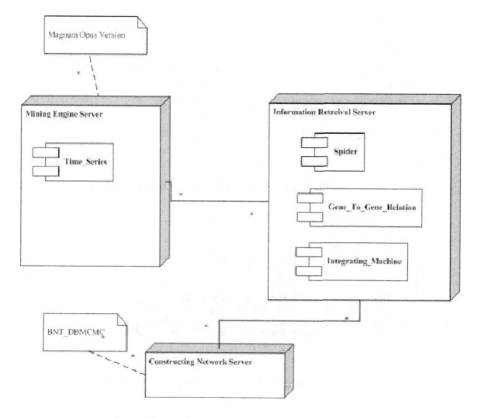

Fig. 7. System deployment diagram

5   Data Source and Finding

We used data from [4] who analyses the cell cycle of Saccharomyces cerevisiae yeast, which contains 6183 genes. We selected a confirmed gene relation (pathway) from KEGG as benchmark that contains 27 genes. A total of 4526 related literature from PubMed was downloaded for further use.

Figure 8 and Figure 9 show the rule scores of gene relations based on the mining engine and retrieval modules without and with data expansion, respectively. The 2 figures show approximately 30,000 rules generated by the mining engine server and sorted by Leverage [22]. They show that our proposed method, which uses all possible time slices to expand the microarray data, can find a higher number of related genes than traditional methods. The scores of most of the rules by the traditional method (Figure 8, without expanded data) are about 100, but those with the proposed method (Figure 9, using all possible time slices to expand the data) reach 200.

Moreover, use of Leverage did not reveal the highest score at the first-order rule. Thus, either the rule at the top is not correct or has less focus in the literature. However, yeast is the one of the widely researched organisms. The rule

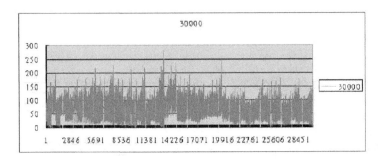

Fig. 8. Traditional association rule without expansion data

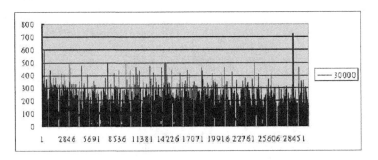

Fig. 9. Expansion data association rule weight value

might not be precise. We therefore suggest that the rule order should be sorted on the basis of both literature and association-rule confidence instead of association-rule score. An algorithm used in the proposed system is described below. The proposed system sets 2 thresholds, α and β, to select suitable rules. To find suitable rules, the user can select a preference of about α and β to emphasize the association rule or literature (0.5 is used for both α and β as the default value). The final order is sent to the constructing network module and a network is built.

If $Rule\_Confidence > \alpha$ and $Rule\_Gene\_Weight > \beta$,
then $Rule\_Weight = 1 - (1 - (Rule\_Confidence)^2) \times (1 - (Rule\_Gene\_Weight)^2)$,
else if $Rule\_Confidence < \alpha'$ and $Rule\_Gene\_Weight < \beta'$,
then $Rule\_Weight = (Rule\_Confidence)^2 \times (Rule\_Gene\_Weight)^2$,
else $Rule\_Weight = Rule\_Confidence \times Rule\_Gene\_Weight$.

6 Conclusion

This article presents a way to use both gene-association rules and literature to find candidate genes for constructing a gene network. The proposed method aims to find the most suitable rule to cluster and reduce the gene comparison number. Use of gene-association rule allows for clustering genes. Accessing the literature to reconfirm the suitability allows for selecting the most suitable rule. Finally,

DBN is used to construct a gene network. With the proposed method, which combines both an association rule and confirmation with the literature, researchers should be able to find positive as well as negative relations more precisely.

Acknowledgement

The research was funded by NSC 93-2213-E-006-064-project from National Science Council Taiwan. We would like to thank the Taiwan Orchid research team for their supports.

References

1. Becquet, C., Blachon, S., Jeudy, B., Boulicaut, J.F. and Gandrillon, O. (2003). Strong-association-rule mining for large-scale gene-expression data analysis: a case study o human SAGE data. *Genome Biol*, 12, pp.1-16.
2. Creighton, C. and Hanash, S. (2003). Mining gene expression databases for association rules. *Bioinformatics*, 19, pp.79-86.
3. Doddi, S., Marathe, A., Ravi, S.S. and Torney, D.C. (2001). Discovery of association rules in medical data. *Med Inform Internet Med*, 26, pp.25-33.
4. Eisen, M.B., Spellman, P.T., Brown, P.O. and Botstein, D. (1998). Cluster analysis and display of genome-wide expression patterns. *Proc. Natl. Acad. Science (USA)*, 95, pp.14863-14868.
5. Eisenberg, D., Marcotte, M.E., Xenarios, I. and Yeates, O.T. (2000). Protein function in the post-genomic era. *Nature*, 405, pp.823-826.
6. Ewing, B. and Green, P. (2000). Analysis of expressed sequence tags indicates 35,000 human genes. *Nature Genet*, 25, pp.232-234.
7. Hieter, P. and Boguski, M. (1997). Functional genomics: its all how you read it. *Science*, 278, pp.601-602.
8. Jenssen, T., Lagreid, A., Komorowski, J. and Hovig, E. (2001). A literature network of human genes for high-throughput analysis of gene expression. *Nature genetics*, 28, pp.21-28.
9. Ji, L. and Tan, K.L. (2004). Mining Gene expression data for positive and negative co-regulated gene cluster. *Bioinformatics*, 20(16), pp.2711-2718
10. Kim, S.Y., Imoto, S. and Miyano, S. (2003). Inferring gene networks from time series microarray data using Dynamic Bayesian Networks. *Briefing Bioinformatics*, 4(3), pp.228-235.
11. Kim, S.Y., Imoto, S. and Miyano, S. (2004). Dynamic Bayesian networks and nonparametric regression for nonlinear modeling of gene networks from time series gene expression data. *Biosystems*, 75, pp.57-65.
12. Murphy, K. and Mian, S. (1999). Modeling gene expression data using dynamic Bayesian networks. Technical Report, Computer Science Division, University of California, Berkeley, CA.
13. Narayanasamy, V., Mukhopadhyay, S., Palakal, M. and Potter, D.A. (2004). TransMiner:Mining Transitive Associations among Biological Objects form Text. *Journal of biomedical science*, 11, pp.864-873.
14. Ong, I.M., Glasner, J.D. and Page, D. (2002). Modeling regulatory pathways in E.coli from time series expression profiles. *Bioinformatics*, 18, pp.241-248.
15. Salton, G., Wong, A. and Yang Cornel, C.S. (1975). A Vector Space Model for Automated Indexing. *Journal of the ACM*, 18(1), pp.613-620.

16. Shatkay, H., Edwards, S. and Boguski, M. (2002). Information retrieval meets gene analysis. *IEEE Intelligent Systems, Special Issue on Intelligent Systems in Biology*, 17(2), pp.45-53.

17. Stephens, M., Palakal, M., Mukhopadhyay, S., Raje, R. and Mostafa, J. (2001). Detecting gene relations from medline abstracts. *Pac Symp Biocomput*, pp. 483-495.

18. Tamayo, P., Slonim, D., Mesirov, J., Zhu, Q., Kitareewan, S., Dmitrovsky, E., Lander, E.S. and Golub, T.R. (1999). Interpreting patterns of gene expression with self-organizing maps methods and application to hematopoietic differentiation. *Nature Genetics*, 96, pp.2907-2912.

19. Tao, Y.C. and Leibel, R.L. (2002). Identifying functional relationships among human genes by systematic analysis of biological literature. *BMC Bioinformatics*, 2002, 3(16), pp.1-9.

20. Tavazoie, S., Hughes, J.D., Campbell, M.J., Cho, R.J. and Church, G.M. (1999). Systematic determination of genetic network architecture. *Nature Genetics*, 22, pp.281-285.

21. Torgeir, R.H., Astrid, L. and Jan, K. (2002). Learning rule-based models of biological process from gene expression time profiles using Gene Ontology. *Bioinformatics*, 19, pp.1116-1123.

22. Webb, G.I. and Zhang, S. (2005). K-Optimal Rule Discovery. *Data mining and Knowledge Discovery*, 10(1), pp.39-79.

23. Zou, M. and Conzen, S.D. (2004). A new dynamic Bayesian network approach for identifying gene regulatory networks from time course microarray data. *Bioinformatics*, Advance Access published on August 12, 2004, pp.1-29.

A Multi-information Based Gene Scoring Method for Analysis of Gene Expression Data

Hsieh-Hui Yu [1], Vincent S. Tseng [1], and Jiin-Haur Chuang [2]

[1] Department of Computer Science and Information Engineering, National Cheng Kung University, Tainan, Taiwan
tsengsm@mail.ncku.edu.tw
[2] Department of Surgery and Internal Medicine, Chang Gung Memorial Hospital at Kaohsiung, Kaoshiung, Taiwan
chou1970@adm.cgmh.org.tw

Abstract. Hepatitis B virus (HBV) infection is a worldwide health problem, with more than 1 million people died from liver cirrhosis and hepatocellular carcinoma (HCC) each year. HBV infection could result in the progression from normal to serious cirrhosis which is insidious and asymptomatic in most of the cases. The recent development of DNA microarray technology provides biomedical researchers with a molecular sight to observe thousands of genes simultaneously. How to efficiently extract useful information from these large-scale gene expression data is an important issue. Although there exist a number of interesting researches on this issue, they used to deploy some complicated statistical hypotheses. In this paper, we propose a multi-information-based methodology to score genes based on the microarray expressions. The concept of multi-information here is to combine different scoring functions in different tiers for analyzing gene expressions. The proposed methods can rank the genes according to the degree of relevance to the targeted diseases so as to form a precise prediction base. The experimental results show that our approach delivers accurate prediction through the assessment of QRT-PRC results.

1 Introduction

Hepatitis B virus (HBV) infection is a serious worldwide health problem, with more than 1 million people died from liver cirrhosis and hepatocellular carcinoma (HCC) each year. Decompensated liver cirrhosis has caused more than half of the death. Although gender, the elders, ongoing HBV replication, HBV genotypes, and concurrent HCV or human immunodeficiency virus infection are predisposing factors, the progression from normal to serious cirrhosis is insidious and asymptomatic in most of the cases. Numerous genes for cytokines, inflammatory and anti-inflammatory regulations, mitochondrial functions, morphogenesis, and apoptosis responses are likely involved in the process of HBV-cirrhosis.

The recent development of DNA microarray technology enables biomedical researchers to simultaneously analyze thousands of genes expressed differently in various clinical conditions. Such gene expression profiles are used to understand the molecular variations among diseases. It is also helpful for medical usages like in developing

C. Priami et al. (Eds.): Trans. on Comput. Syst. Biol. V, LNBI 4070, pp. 97–111, 2006.

diagnostic tools. How to efficiently and effectively extract useful information from these large-scale gene expression data is an important issue. In recent years, a number of data mining methods are proposed, like clustering [2,6,12,15,22], classification [5,10,14,18,20,23], association analysis, evolution analysis and so on [1]. In this study, we focus on the classification area for finding the most informative genes related to particular classes of diseases. However, the large number of genes compared to small samples results in difficulties of back-end analysis. Since the researchers must precisely extract informative genes from several samples, it will be very likely to make some fatal errors if non-adaptive analysis methods are used. In many literatures, we found that different analysis methods fit for different types of data sets [3, 4, 7, 8, 9, 17, 19, 25, 26]. There is hardly a method which is generally suitable for all kinds of data sets. Hence, it is indeed a critical issue to select an adaptive method.

In this study, we proposed a "multi-information based scoring methodology" to score genes based on the microarray expressions, with the aim to find out interesting genes related to targeted diseases. The concept of multi-information is to combine different scoring functions in different tiers, so that we can get a ranked value for each gene through multiple evaluations. The main advantages of the proposed methodology include: 1) Simplicity: Our methodology combines different statistical scoring functions. It doesn't adopt any complicated criteria but still achieve good results. 2) High flexibility: We can adjust the combination of scoring functions according to the types of input data. 3) Noise filtering: By the combination of some scoring functions, the noisy data could be filter out in each tier so that more accurate results could be obtained. 4) High precision: From the experimental results, our methodology is shown to have high prediction ability in finding out interesting genes related to targeted diseases. In this research, the experimental results on analysis of liver cirrhosis and hepatocellular carcinoma (HCC) show that our approach has accurate prediction through the assessment of QRT-PRC results [13, 21].

The remainder of this paper is organized as follows. In section 2, we describe the framework and related work for the proposed methodology. In section 3, the experimental results are presented to assent our method. We conclude and describe the future work in section 4.

2   Proposed Methodology

Figure 1 shows the flowchart of the proposed methodology. The system consists of several modules, namely preprocessing module, data cleaning module, multi-info based gene scoring module, validation module and the produced informative genes database. The aim of the proposed methodology is to provide a standard model for discovering informative genes by analyzing gene expression data. As discussed in Section 1, since there is no single perfect scoring method fitted for all kinds of data in discovering useful information, we try to combine different kinds of scoring methods in different orders. Each step in the multi-info based gene scoring module can filter out some noisy data or divide the whole data into different groups. Different scoring methods could be applied to different groups of data. Through the combination of some scoring methods, we can extract informative gene with high relevance to the analysis targets. In this study, we test the TNoM method proposed by Ben-Dor *et al.* [4] and

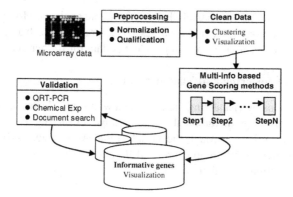

Fig. 1. Flowchart of proposed methodology for informative genes extraction

Gaussian overlap method [27] in filtering genes. Certainly, we could use more kinds of analysis methods in the methodology to extract informative gene with high relevance to the analysis targets from amounts of gene expression data effectively. In the following, we describe the each module in details.

2.1 Preprocessing

The raw gene expression data may contain many errors resulted during the experimental processes [11,16]. For example, it may make different results between arrays when the machine spots RNA into the microarray chips or washes chips. We need to perform preprocessing actions to eliminate these biased errors. Two main actions are used here, namely normalization and qualification.

2.1.1 Normalization

First, we should consider the biased differences among distinct experimental results. To reduce the differences made during the experiments, we must normalize the data under some norms to achieve it. For example, we can simply normalize each microarray chip by dividing the mean of the chips so that the differences incurred in different experiments can be reduced. Moreover, we may also normalize the gene expression value to a specified range. Alternately, more complex methods can be adopted to normalize the data according to analysis targets for advanced analysis.

2.1.2 Qualification

During the experimental procedures, some erroneous results could be produced. We must exclude these data to get more accurate analysis results. Besides, the fact that each gene is normally duplicated on one array chip should also be considered. In theory, the same genes of a patient in one chip should have similar expression values. Thus the variation between the replicate spots can be used to assess the reliability of the gene measurement. Informally, if the replicated measurements are close to each other, then the estimation of the gene expression can be obtained by a combination (e.g., average) of the replicates. On the other hand, if the replicates are quite different, then the gene will be flagged as unreliable and excluded in subsequent analysis. Moreover, we should exclude genes that express negative values by subtracting the backgrounds values. Through the procedures above, we can get cleaner data for further analysis.

2.2 Clean Data Clustering and Visualization

We may gain insights by visualizing the clean data with some clustering methods. In our experiment, we use the hierarchical agglomerative clustering (HAC) methods to investigate the clean data. Through the visualization, we can get useful observations, e.g., some genes are very differential among different classes. This kind of visualization can help the conduction of further analysis.

2.3 Multi-information Based Gene-Scoring Method

Since the difficulty and expense of back-end analysis of tremendous amount of data preclude the inclusion of large number of subjects in most studies, a reliable and effective process is needed to extract genes of high relevance to the analysis target. Since there is no perfect method fitted for all kinds of data, we proposed a multi-info based gene scoring methodology to achieve this goal. The basic idea of the multi-info based gene scoring methodology is to combine different scoring functions to extract informative genes. Hence, we don't need to deploy some complicated mathematical functions in analyzing data. Instead, we combine some known scoring methods in different tiers. In the first run, we use the first-tier scoring method to classify all genes into different classes according to their expressions. Then, the next-tier scoring method is applied on the result obtained from previous pass. Finally, the final-tier scoring method is applied to obtain final ranked scores for all genes. Through these processes, we can get Top N informative genes as we wanted from the sorted result. Every tier in the multi-info based gene scoring methods module can filter out some noisy data or divide data into different groups. In this way, we can extract informative gene with high relevance to the analysis targets.

In this study, we adopted two scoring methods, namely the *threshold number of misclassification* (TNoM) [4] and the *Gaussian overlap* [27], to form a multi-info based gene scoring methodology to identify genes related to HBV-cirrhosis. We apply the TNoM and the Gaussian overlap in order. The details of the two scoring methods are described in the following.

2.3.1 Threshold Number of Misclassification (TNoM)

We now briefly describe the scoring method TNoM. Assume that tissue A is labeled as "+" (positive sample), and B labeled as "−" (negative sample). Let g be a gene that we want to score for a partition between positive and negative samples. Intuitively, for the gene g, there will be one part relevant to the other that is over-expressed or vice-verse. To formalize the notion of relevance between the negative and positive, we sort the tissues according to the expression levels of gene g. Let $\pi = <\pi_1,..., \pi_{a+b}>$ be the permutation of the tissues induced by the expression value of gene g sorted by ascending order. The rank vector v of g is a $\{+,-\}$ vector of length a+b, where v_i is the label of tissue π_i.

For example, if the expression value of g in positive tissue samples are {10, 20, 40, 80}, and {70, 85, 90, 140} in negative tissue samples, then

π = <10, 20, 40, 70, 80, 85, 90, 140> and
v = <+, +, +, −, +, −, −, −>

Intuitively, the rank vector v profiles the essence of differential expression value of gene g. If g is under-expressed in one class, then the tissue samples of this class will be concentrated at the left side of the rank vector v, and the other one will be concentrated at the right side. So, we know that, if gene g is informative enough, then most of the positive class will be concentrated at one side, and the negative class will be in the other side. The TNoM scores, described below are two natural ways to score a rank vector based on its most homogeneous partition. Formally, the TNoM score of a vector v in a gene g is defined as:

$$TNoM(v) = \min_{x; y = v} \min([\#_-(x) + \#_+(y)], [\#_+(x) + \#_-(y)]) \qquad (1)$$

where $\#_s(x)$ is the number of times a symbol s (+ or -) appears in the vector v. Thus, for each partition x and y of v, we first consider the label x as positive and y as negative. In this case, we count $\#-(x) + \#+(y)$. Then, we consider the opposite by counting $\#+(x) +\#-(y)$. Finally, we return the partition for which the best classification makes the smaller number of misclassification.

For example, for the rank v above, the best partition of v into two parts is

v = <+, +, +, −, +> ;<−, −, −>,

Thus, TNoM (v) = 1 + 0 = 1. Note that the partition of v is equivalent to choosing a threshold for expression value and counting the misclassification number. From the TNoM score, a gene is considered as more informative if the misclassification number is smaller.

2.3.2 Gaussian Overlap
Gaussian overlap is a method to calculate distribution overlap between expression values for genes in different classes. The score is based on normality assumption. T smaller the score, more informative the gene is.

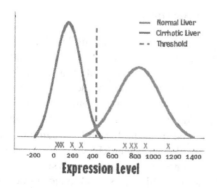

Fig. 2. Gaussian Overlap of a gene between normal liver and cirrhotic liver

For example, Figure 2 shows the Gaussian distribution of a gene in two different classes of liver tissues. The overlap of the two distribution represents the probability of a gene expresses similarly in the two different classes. The smaller the Gaussian overlap vale, the more powerful a gene is in discriminating the two classes. In the paper, we utilize this property to identify the difference of diseases.

2.4 Interesting Gene Clustering and Visualization

Through the processes proposed above, the Top N informative genes could be obtained, which represent the interesting genes for the analysis targets. We may visualize these top N formative genes through clustering methods. In this study, we use the hierarchical agglomerative clustering (HAC) method to examine the interesting genes. From the visualization of interesting genes versus clean data, we can easily find that the interesting genes are really informative than other ones. Meanwhile, we can also validate the correctness of the analysis methods.

2.5 Validation

Through the analysis modules, we can obtain the informative genes. However, we are not sure whether the obtained informative genes are correct or not. Hence, we need some validation methods to validate the result. As we know, some chemical experiments like RT-PCR can validate the result correctly. Besides, we can also do the validation through literature studies. From these steps, the informative genes we collect will be more reliable. Here, we adopt the RT-PCR to verify the analysis results. Subsequently, the reliable interesting genes assessed by RT-PCR can be gathered together to form a prediction model for the diseases. In this paper, RT-PCR is performed by using ABI 7700 Sequence Detection System (TaqMan, Perkin-Elmer Applied Biosystems), to confirm analysis results from oligonucleotide micoarray. PCR was performed in 50 µl SYBR Green PCR Master Mix (Applied Biosystems) containing 10 µM forward primers and reverse primers, and approximately 30 ng cDNA.

3 Experimental Results

3.1 Datasets for Liver Specimens

For microarray analysis, we use five liver samples. Three are diseased samples which were obtained from the liver explants of the adult patients with decompensated HBV-cirrhosis at the time of liver transplantation. The other two samples are control liver ones which were obtained from the two adult patients receiving liver resection for focal nodular hyperplasia in one and for hemangioma in the other. The liver samples were snap-frozen in liquid nitrogen and stored at $-80°C$ until use.

Total RNA were isolated from the frozen liver samples using a single-step method with the commercially available reagent REZOL^{TM}C&T (Protech technology, Taiwan). Then we sent all RNA data to microarray analysis. Gene expression profiles were obtained using Affymetrix [28] U133A oligonucleotide arrays [26]. The Human Genome U133A array consists of 22,283 probe sets annotated by 13,624 UniGene [30] and 12,769 LocusLink [29] identifiers. The study was approved by the Institutional Review Board of the Chang Gung Memorial Hospital at Kaohsiung, Taiwan.

3.2 Analysis Results

For the five oligonucleotide arrays, each of them contains 22,283 genes. For the raw data, we need to preprocess them for later analysis. First, we eliminate the genes with missing values in the five samples. Thus we get 6,691 genes with complete expression values in the five samples from 22,283 genes. Then, we normalize each gene with the mean of the 6,691 genes in each chip. The expression pattern was shown in Figure 3 by

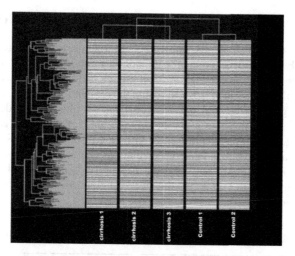

Fig. 3. Hierarchical Agglomerative Clustering of 6691 genes from 5 specimens

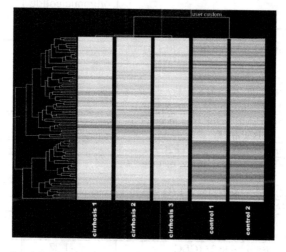

Fig. 4. Hierarchical Agglomerative Clustering of the top 100 genes from 5 specimens

the Hierarchical Agglomerative Clustering. As we can see, the difference between cirrhotic and normal liver is not obvious. It's difficult to identify the disease by these genes. To obtain informative genes, we input these genes into the multi-information based gene scoring module. Here, we combine two scoring functions: TNoM and Gaussian overlap measurement. First, the parameter of TNoM was set to zero. Consequently, we get 3,203 genes from the 6,691 genes. Then, we use the Gaussian overlap to rank these genes and chose the top 100 genes for further analysis. As Figure 4 shows, there exist obviously differential patterns between cirrhosis and normal samples. Besides, two distinct groups of patterns were identified. One is up-regulated pattern with 37 genes and the other is down-regulated with 63 genes by comparing cirrhosis to normal samples.

3.3   Correlation of the Fold Change in Oligonucleotide Array with QRT-PCR

We randomly selected 6 genes that were up-regulated and 6 genes that were down-regulated in cirrhosis (compared to normal) for verification with the use of QRT-PCR. Four out of 6 in each up or down groups met the usual selection criteria of greater than 2-fold, or less than 0.5-fold, while the remaining two were in-between 0.5- to 2-fold. The gene name and the fold changes in oligonucleotide array and in QRT-PCR were shown in Table 1. As shown in Figure 5, there was a good correlation of the fold changes in oligonucleotide array with the fold changes in QRT-PCR with $r^2 = 0.903(r = 0.950)$ and P=0.000 in Pearson correlation. The fold change of the 12 genes in QRT-PCR was further confirmed with the use of 1.5% agarose gel electrophoresis (Tab. Y1). From the QRT-PCR results, we can find out that our methodology is precisely and practical.

Table 1. The gene name and the fold changes in oligonucleotide array and in QRT-PCR

Gene Name	Fold changes in oligonucleotide array	Fold changes in QRT-PCR
Annexin A2	4.14	4.09
Actin binding LIM protein 1 (ABLIM)	4.02	2.41
Sushi-repeat-containing protein, X chromosome (SRPX)	2.84	1.97
Follistatin-like 1 (FSTL1)	2.09	1.45
Ras-related GTP-binding protein (RAGA)	1.67	1.31
Transforming growth factor, beta receptor 3 (TGF-βR3)	1.60	1.11
Cytochrome P-450, 2C18 (CYP2C18)	-8.43	-13.89
Tumor necrosis factor superfamily, member 10 (TNFSF10)	-3.60	-12.99
Death-associated protein kinase 1 (DAPK1)	-2.34	-5.88
Aquaporin 9 (AQP9)	-2.02	-1.97
Monoamine oxidase B, nuclear gene encoding mitochondrial protein (MAOB)	-1.91	-1.69
Cell division cycle 23 (CDC23)	-1.27	-1.31

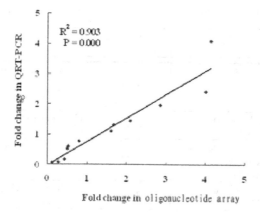

Fig. 5. The correlation of the fold changes between oligonucleotide array and QRT-PCR

3.4   Characteristics of 13 Up-Regulated Genes in Cirrhotic Liver

There were 37 genes up-regulated in HBV-cirrhosis, in which 13 genes of known function increased for more than 2-fold over the normal control liver. The list of these 13 genes, including fold change over noncirrhotic liver on the oligonucleotide microarray and their individual function was shown in Table 2. On the top of the list was Annexin A2 (ANXA2), also known as phospholipase A2, which was a calcium-dependent phospholipid binding protein important in cell growth and signal transduction. Up-regulation of this gene in liver cirrhosis might compensate apoptosis of the liver cells induced by other genes during liver cirrhosis. The second one was actin-binding LIM protein (ABLIM), which was implicated in cytoskeletal organization and in electron transport. Up-regulation of this gene implied either active cytoskeletal remodeling in liver cirrhosis or a compensatory reaction to retarded electron transport. Next to ABLIM was protein kinase C and casein kinase substrate in neuron 2 (PACSIN2). PACSIN2 was a regulator of metalloprotease ADAM13 and it interacted with CD95L to induce apoptosis or for T-cell activation. This gene was obviously involved in morphogenesis and possibly fibrogenesis characteristic of liver cirrhosis. The other 10 genes functioned as molecular chaperone (HSPA4), cation transport (ATP1A1), translation initiation (BZW2), coregulation of other gene (SRPX), ATP binding and protein folding (HSPA8), cytokine (GS 3786), vesicle transport (CKAP4), actin modulation (CFL1), inflammation (FSTL1) and alcohol metabolism (APA1). Except PACSIN2, none of other genes were strictly defined as morphogenetic or fibrogenic. The up-regulation of these genes might reflect an active repair mechanism still existed even in the end stage of liver cirrhosis.

3.5   Characteristics of 23 Genes Down-Regulated in Cirrhotic Liver

There were 63 genes down-regulated in cirrhosis in which 23 genes of known function decreased for more than 2-fold over the control. A list of these genes was shown in Table 3. Interestingly, 11 genes were involved in metabolism of steroids, lipid or proteins, or in detoxifying carcinogens and xenobiotics. Down-regulation of these

Table 2. Function and fold change of 13 genes up-regulated in HBV-cirrhosis

Gene Name	Fold change	Function
Annexin A2	4.14	Phospholipid binding protein; cell growth
		Signal transduction
Actin binding LIM protein 1 (ABLIM)	4.02	Cytoskeleton organization / electron transporter
Protein kinase C and casein kinase substrate in neurons 2 (PACSIN2)	3.77	DNA binding
Heat shock protein apg-2 (HSPA4)	3.19	Molecular chaperones / protein folding, translocation
ATPase, Na+K+ transporting, alpha 1 polypeptide (ATP1A1)	2.96	Metabolism / cation transporter
Homo sapiens basic leucine zipper and W2 domains 2 (BZW2)	2.89	Translation initiation factor
Sushi-repeat-containing protein, X chromosome (SRPX)	2.84	Coregulation of neighbour gene
Constitutive heat shock protein 70 (HSPA8)	2.69	ATP binding / protein folding
Predicted osteoblast protein (GS3786)	2.33	Cytokine
Homo sapiens cytoskeleton-associated protein 4 (CKAP4)	2.22	Non-selective vesicle transport
Cofilin 1 (CFL1)	2.21	Actin modulating / actin binding
Follistatin-like 1 (FSTL1)	2.09	Heparin and calcium binding / inflammation
Another partner for ARF 1 (APA1)	2.04	Zinc-dependent alcohol dehydrogenase

genes implied impairment of the function in the involved metabolic pathway. On the top of the list was P450, 2C18 (CYP2C18), which was one of the four CYP2C isoforms. The cytochrome P450 proteins were monooxygenases, which catalyzed many reactions involved in drug metabolism and synthesis of cholesterol, steroids and other lipids. Next to CYP2C18 was gamma-butyrobetaine hydroxylase (BBOX1). This gene encoded gamma butyrobetaine hydroxylase, which catalyzed the formation of L-carnitine from gamma butyrobetaine and was therefore involved in mitochondrial beta-oxidation. The other 9 genes of metabolism included folate hydrolase 1 (FOLH1), aminolevulinate delta dehydratase (ALAD), solute carrier family 38 member4 (SLC38A4), isocitrate dehydrogenase 1 (IDH1), arylamine N-acetyltransferase 1 (NAT1), soluble epoxide hydroxylase (EPHX2), thioesterase superfamily member 2 (THEM2), aldehyde dehydrogenase 3 family member A2 (ALDH3A2), and dopa decarboxylase (DDC). The function of some of these genes, including FOLH1, IDH1, THEM1, ALDH3A2 and DDC was obvious from the name, but the function of 4 other genes was not. For example, ALAD was a porphobilinogen synthase in heme biosynthesis. A defect of this gene might increase sensitivity in lead poisoning and

Table 3. Function and fold change of 23 genes down-regulated in HBV-cirrhosis

Gene Name	Fold change	Function
Cytochrome P450, 2C18 (CYP2C18)	-8.44	Metabolism / mmonooxygenase / electron transport
Gamma-butyrobetaine hydroxylase (BBOX1)	-5.52	Metabolism / mitochondrial beta-oxidation
Cytoplasmic polyadenylation element binding protein 3 (KIAA 0940)	-4.68	Nucleic acid binding
Folate hydrolase 1(FOLH1)	-3.91	Metabolism / proteolysis / peptidolysis
Aminolevulinate, delta-, dehydratase	-3.87	Metabolism / heme biosynthesis
Growth hormone receptor (GHR)	-3.69	Growth hormone receptor
Tumor necrosis factor (ligand) superfamily, member 10 (TNFSF10)	-3.60	Cytokine
Solute carrier family 38, member 4 (SLC38A4)	-3.46	Metabolism / amino acid transport
Nuclear receptor binding SET domain protein 1 (NAD1)	-2.95	Androgen receptor coregulator
Isocitrate dehydrogenase 1 (IDH1)	-2.91	Metabolism / NADPH production
Arylamine N-acetyltransferase1 (NAT1)	-2.89	Metabolism; detoxify carcinogen; xenobiotics
Annexin 14 (ANX14)	-2.75	Phospholipid and calcium binding / anticoagulant
Soluble epoxide hydrolase (EPHX2)	-2.63	Metabolism / epoxide hydrolase
A kinase anchor protein 9 (AKAP9)	-2.56	Small molecule transport / signal transduction
Death-associated protein kinase 1 (DAPK1)	-2.34	Protein amino acid phosphorylation; signal transduction; ATP binding ; tumor suppressor
Thioesterase superfamily member 2 (THEM2)	-2.27	Enzyme
Cell division cycle 2-like 2 (CDC2L2)	-2.25	Protein amino acid phosphorylation; ATP binding; tumor suppressor
Checkpoint suppressor 1 (CHES1)	-2.18	DNA damage checkpoint /regulation of transcription
Aldehyde dehydrogenase 3 family, member A2 (ALDH3A2)	-2.18	Epidermal differentiation; nervous system development; lipid metabolism; detoxification
Dopa decarboxylase (DDC)	-2.15	Metabolism / amino acid metabolism
Structural maintenance of chromosomes 6-like 1 (SMC6L1)	-2.13	Chromosome segregation / ATP binding
Suppression of tumorigenicity 13 (ST13)	-2.08	Adaptor protein for HSP70 and HSP90; tumor suppressor; glucocorticoid receptor assembly
Aquaporin 9 (AQP9)	-2.02	Water channel transporter

caused acute hepatic porphyria. SLC38A4 was actually an amino acid transporter. NAT1 had acetyltransferase activity and functioned to detoxify carcinogens and xenobiotics. EPHX2 was a hydrolase, also involving in drug metabolism and oxidative stress.

Next to metabolism were three genes functioning as cell signaling, which included cytoplasmic polyadenylation element binding protein 3 (KIAA0940), A-kinase anchor protein 9 (AKAP9) and death-associated protein kinase 1 (DAPK1). KIAA0940 gene encoded a protein implicated in cytoplasmic polyadenylation and translational regulations for many growth factors. AKAP9 encoded a member of AKAP proteins that functioned to bind to the regulatory subunit of protein kinase A and interacted with numerous signaling proteins. DAPK1 also had protein kinase activity and involved in signal transduction and protein amino acid phosphorylation and was a tumor suppressor candidate. Down-regulation of the latter might imply loss of tumor suppressor function and facilitated the development of hepatocellular carinoma characteristic of HBV-cirrhosis.

Three genes that involved in DNA damage checkpoint, chromosome segregation or tumor suppression were down-regulated. Checkpoint suppressor 1 CHES1), as its name implied, was a member of the forkhead/winged helix transcription factor family that suppressed multiple yeast check point mutations. This gene was therefore important in DNA damage checkpoint and in regulation of transcription. Structural maintenance of chromosome 6-like 1 (SMC6L1) encoded a structural maintenance of chromosome protein SMC, C-terminal family member and was involved in chromosome segregation. Suppression of tumorigenicity 13 (ST13) encoded an adaptor protein that mediated the association of the heat shock protein HSP70 and HSP90. Down-regulation of this gene in the colorectal cancer tissue implied its role as candidate tumor suppressor gene. As a whole, decreased expression of the above three genes in liver cirrhosis might brood a milieu in which HCC was more easily to develop.

There were two genes functioned as nuclear receptor or coactivator, including growth hormone receptor (GHR) and nuclear receptor binding SET domain protein 1 (NSD1). As its name implied, GH bound to GHR and activated signal transduction pathway that resulted in mitogenic and anabolic response leading to growth. NSD1 was an androgen receptor coactivator. Down-regulation of both genes might imply decompensation of anabolism characteristic of liver cirrhosis.

Three other genes that were down-regulated in HBV-liver cirrhosis were tumor necrosis factor superfamily, member 10 (TNFSF10), annexin14 (ANX14), and aquaporin 9 (AQP9). TNFSF10 was a cytokine that belonged to the TNF ligand family that induced apoptosis and involved in immune response. Down-regulation of TNFSF10 might also be deleterious to the cirrhotic liver for its potential tumorigenesis. ANX14 (also known as ANXA10), although functioned like ANXA2 in calcium-dependent phospholipid binding, were downregulated in HBV-cirrhosis. Such a discordant finding implied diverse biological processes in the annexin family and would be discussed later. AQP9 belonged to a family of water-selective membrane channels, but assumed broad solute permeability. Down-regulation of this gene in cirrhosis not only indicated problem in water transport, but also implied impairment of glycerol and urea turnover.

4   Conclusions

In this study, we have proposed a multi-information-based methodology for scoring genes based on the microarray expression values. The proposed approach contains a complete analysis flow including several modules, namely preprocessing module, data cleaning module, multi-information based gene scoring module, and validation module. In the experiments, we combine TNoM and Gaussian overlap methods in different tiers to analyze liver cirrhosis and hepatocellular carcinoma. From the clustering results, we can see that there are two distinct groups of gene expression patterns between cirrhosis and normal tissues. Some of these genes have been known to relate to liver cirrhosis. But, there are still many genes with known functions not being found to be related to liver cirrhosis, which are worth of deeper analysis. The QRT-PCR validation result also shows that our methodology is precise and has high predictive power in finding out interesting genes related to targeted diseases.

In the future, we shall go for some extended research work. For the scoring methods, we shall test more kinds of combinations with different gene scoring methods. By recording the effects and characteristics of these combinations according to different type of genes, the system may automatically select suitable scoring functions according to different type of data. For general applications, we shall build up an on-line prediction platform by integrating useful biological databases like pathway and protein information for deeper analysis.

Acknowledgement

This research was partially supported by National Science Council, Taiwan, R.O.C., with grant no. NSC 94-2213-E-006 -099.

References

1. Alizadeh, A. A., Eisen, M. B., Davis, R. E., Ma, C., Lossos, I. S., Rosenwald, A., Boldrick, J. C., Sabet, H., Tran, T., Yu, X., Powell, J. I., Yang, L., Marti, G. E., Moore, T., Hudson, J., J., Lu, L., Lewis, D. B., Tibshirani, R., Sherlock, G., Chan, W. C., Greiner, T. C., Weisenburger, D. D., Armitage, J. O., Warnke, R., Staudt, L. M. et al., 'Distinct types of diffuse large B-cell lymphoma identified by gene expression profiling', Nature 403(6769), 503–11, 2000.
2. Alon, U. et al., 'Broad patterns of gene expression revealed by clustering analysis of tumor and normal colon tissues probed by oligonucleotide arrays', Proceedings of the National Academy of Sciences, vol. 96, pp. 6745-6750, 1999.
3. Ben-Dor, A., Friedman, N. and Yakhini, Z., 'Scoring genes for relevance', Technical Report 2000-38, School of Computer Science & Engineering, Hebrew University, Jerusalem.
4. Ben-Dor, A., Friedman, N. and Yakhini, Z., 'Overabundance Analysis and Class Discovery in Gene Expression Data', Technical Reports of the Leibniz Center, 2002.
5. Ben-Dor, A., Bruhn, L., Friedman, N., Nachman, I., Schummer, M. and Yakhini, Z. 'Tissue classification with gene expression profiles', Jour. Of Comp. Bio., 7: 559-584, 2000.
6. Ben-Dor, A., Shamir, R. and Yakhini, Z., 'Clustering gene expression patterns', J. Comp. Bio. 6 (3-4), 281–97, 1999.

7. Blum, A. and Langley, P., 'Selection of relevant features and examples in machine learning.', Artificial Intelligence, vol. 97, pp. 245-271, 1997.

8. Chuang, H. Y., Liu, H. F., Brown, S., Cameron, M. C. and Kao, C. Y., 'Identifying significant genes from microarray data', fourth IEEE Symposium on Bioinformatics and Bioengineering (BIBE), 358-366, 2004.

9. Chuang, H. Y., Tsai, H. K., Tsai, Y. F. and Kao, C. Y., 'Ranking genes for discriminability on microarray data.', Journal of Information Science and Engineering, vol. 19, pp. 953-966, 2003.

10. Cortes, C. and Vapnik, V., 'Support vector machines', Machine Learning 20, 273–297, 1995.

11. de Kok, J. B., Roelofs, R. W. , Giesendorf, B. A., Pennings, J. L., Waas, E. T., Feuth, T., Swinkels, D. W. and Span, P. N., 'Normalization of gene expression measurements in tumor tissues: comparison of 13 endogenous control genes.', Lab Invest. Jan;85(1):154-9, 2005.

12. Eisen, M. B., Spellman, P. T., Brown, P. O. and Botstein, D., 'Cluster analysis and display of genome-wide expression patterns', PNAS 95(25), 14863–8, 1998.

13. Gerard, C. J., Andrejka, L. M. and Macina, R. A.., 'Mitochondrial ATP synthase 6 as an endogenous control in the quantitative RT-PCR analysis of clinical cancer samples.' Mol Diagn, 5: 39–46, 2000.

14. Golub, T. R., Slonim, D. K., Tamayo, P., Huard, C., Gaasenbeek, M., Mesirov, J. P., Coller, H., Loh, M. L., Downing, J. R., Caligiuri, M. A., Bloomfield, C. D. and Lander, E. S. , 'Molecular classification of cancer: class discovery and class prediction by gene expression monitoring', Science 286(5439), 531–7, 1999.

15. Jain, A. K. and Dubes, R. C., 'Algorithms for Clustering Data', Prentice Hall, 1988.

16. Kunth, K., Hofler, H. and Atkinson, M. J., 'Quantification of messenger RNA expression in tumors: which standard should be used for best RNA normalization?' Verh Dtsch Ges Pathol, 78: 226–230, 1994.

17. Marden, J. I., 'Analysing and Modeling Rank Data.', Chapman & Hall, 1995.

18. McQueen J., 'Some Methods of Classification and Analysis of Multivariate Observations', Proc. of the 5th Berkeley Symp. Mathematical Statistics and Probability, pp. 281-297, 1967.

19. Park, P. J., Pagano, M., and Bonetti, M., 'A Nonparametric Scoring Algorithm for Identifying Informative Genes from Microarray Data.', Pacific Symposium on Biocomputing, vol. 6, pp. 52-63, 2001.

20. Pavlidis, P. and Tang, C., 'Classification of genes using probabilistic models of microarray expression profiles', Proceedings of BIOKDD 2001.

21. Schmittgen, T. D. and Zakrajsek, B.A., 'Effect of experimental treatment on housekeeping gene expression: validation by real-time, quantitative RT-PCR.' J Biochem Biophys Methods, 46: 69–81, 2000.

22. Sharan, R. and Shamir, R., 'CLICK: A clustering algorithm with applications to gene expression analisys', in 'ISMB'00', 2000.

23. Slonim, D. K., Tamayo, P., Mesirov, J. P., Golub, T. R., and Lander, E. S., 'Class prediction and discovery using gene expression data', In RECOMB. 2000.

24. Staunton, J. E., Slonim, D. K., Coller, H. A., Tamayo, P., Angelo, M. J., Park, J., Scherf, U., Lee, J. K., Reinhold, W. O., Weinstein, J. N. et al., 'Chemosensitivity prediction by transcriptional profiling.' Proc Natl Acad Sci USA 2001, 98:10787-10792.

25. Weston, J., Mukherjee, S., Chapelle, O., Pontil, M., Poggio, T. and Vapnik, V., 'Feature selection for SVMs.', In Advances in Neural Information Processing Systems, MIT Press, vol. 13, 2001.

26. Xu, L., Krzyzak, A. and Suen, C.Y., 'Method of Combining Multiple Classifiers and their Application to Handwriting Recognition.', IEEE Trans SMC, vol. 22, pp. 418-435, 1992.

27. Zuo, F., Kaminski, N., Eugui, E., Allard, J., Yakhini, Z., Ben-Dor, A., Lollini, L., Morris, D., Kim, Y., DeLustro, B., et al., 'Gene expression analysis reveals matrilysin as a key regulator of pulmonary fibrosis in mice and humans.' Proc Natl Acad Sci USA 2002;99:6292–6297.
28. Affymetrix. User's guide to product comparison spreadsheets. 2003. http://www.affymetrix.com/support/technical/manual/comparison_spreadsheets_manual.pdf
29. LocusLink: http://www.ncbi.nlm.nih.gov/LocusLink
30. UniGene: http://www.ncbi.nlm.nih.gov/entrez/query.fcgi?db=unigene

Relation-Based Document Retrieval for Biomedical IR[*]

Xiaohua Zhou[1], Xiaohua Hu[1], Guangren Li[2], Xia Lin[1], and Xiaodan Zhang[1]

[1] College of Information Science & Technology, Drexel University
3141 Chestnut Street, Philadelphia, PA 19104
[2] Faculty of Economy, Hunan University, Changsha, China
xiaohua.zhou@drexel.edu, {thu, xlin, xzhang}@cis.drexel.edu

Abstract. In this paper, we explore the use of term relations in information retrieval for precision-focused biomedical literature search. A relation is defined as a pair of two terms which are semantically and syntactically related to each other. Unlike the traditional "bag-of-word" model for documents, our model represents a document by a set of sense-disambiguated terms and their binary relations. Since document level co-occurrence of two terms, in many cases, does not mean this document addresses their relationships, the direct use of relation may improve the precision of very specific search, e.g. *searching documents that mention genes regulated by Smad4*. For this purpose, we develop a generic ontology-based approach to extract terms and their relations, and present a betweenness centrality based approach to rank retrieved documents. A prototyped IR system supporting relation-based search is then built for Medline abstract search. We use this novel IR system to improve the retrieval result of all official runs in TREC-2004 Genomics Track. The experiment shows promising performance of relation-based IR. The average P@100 (the precision of top 100 documents) for 50 topics is significantly raised from 26.37 %(the P@100 of the best run is 42.10%) to 53.69% while the MAP (mean average precision) is kept at an above-average level of 26.59%. The experiment also shows the expressiveness of relations for the representation of information needs, especially in the area of biomedical literature full of various biological relations.

1 Introduction

Precision (the proportion of relevant documents in all retrieved documents) and recall (the proportion of retrieved relevant documents in all relevant documents in the collection) are two basic metrics to measure the performance of Information Retrieval (IR). Often, high precision is at the cost of low recall, and vice versa. Nowadays, precision-focused searching is getting more and more attention most likely due to the following two reasons. First, in a lot of domain-specific application-related search, such as searching the Medline, which collects 14 millions of biomedical abstracts published in more than 4600 journals, the biomedical professional normally know what they need and their search queries are often very specific and only like to receive

[*] This research work is supported in part from the NSF Career grant (NSF IIS 0448023). NSF CCF 0514679 and the research grant from PA Dept of Health.

those documents which meet their specific query, they do not expect a large number of documents. Second, the absolute number of returned relevant document is still large enough for majority of tasks even if the recall is low because of the exponentially increasing size of collections.

Term-based IR models view a document as a bag-of-term, i.e. each term is treated independently without considering the possible connections or relationships. They assign each term a weight by various methods such as TF*IDF family methods [9, 14] and language modeling methods [13] while computing the similarity between document and query. They do not explicitly address the semantics of terms either thought some approaches such as latent semantic indexing [3] try to identify the latent semantic structure between terms. Basically, this line of statistical approaches is efficient and effective in IR. However, they may not be effective to approach very specific information needs that address the relationship between terms.

Term-based IR models have to use term co-occurrence to approximate term relations because there are no direct relations available in their models. However, the co-occurrence of two terms in a document, in many cases, does not mean this document really addresses their relationships, especially when the co-occurrence count is low (e.g. in abstract-based search, the co-occurrence count is often low). Thus, the precision would be compromised. We conducted a simple experiment that tried to retrieve documents addressing the interaction of *obesity* and *hypertension* from PubMed[1] by specifying the co-occurrence of term *hypertension* and *obesity* in abstract or title. We then took the top 100 abstracts for human relevance judgment. Unfortunately, as expected, only 33 of them were relevant.

obesity [TIAB] AND hypertension [TIAB] AND hasabstract [text]
AND ("1900"[PDAT] : "2005/03/08"[PDAT])

Fig. 1. The query used to retrieve documents addressing the interaction of obesity and hypertension from PubMed. A ranked hit list of 6687 documents is returned.

Based on this finding, we develop a precision-focused IR model for domain-specific search, which basically treats a document a set of sense-disambiguated terms and their binary relations. A relation is defined as a pair of two terms which are semantically and syntactically related to each other. Since a relation in our model is explicitly asserted, the direct use of relation in IR may improve the precision of domain-specific search though the recall may be slightly lowered.

Retrieval of biomedical literature often involves various specific biological relations. Take the example of TREC 2004 Genomics Track [7] the goal of which is to study retrieval tasks in genomic domain. All 50 ad hoc retrieval topics[2] are compiled from real information needs of scientists in biomedical domain and most of them are about very specific relationships among gene (protein), mutations, genetic functions, diseases and so on (see some examples in Section 2.2). For this reason, relation-based IR is an appropriate approach to biomedical literature search.

---

[1] http://www.ncbi.nlm.nih.gov/entrez/query.fcgi
[2] http://trec.nist.gov/data/genomics/04.adhoc.topics.txt

The extraction of binary relations from text is a challenging task. We think this is one of the reasons that there is no relation-based approach to IR reported so far. Term extraction is the first step of relation extraction. The methods for term extraction fall into two categories, with dictionary [12, 20] or without dictionary [11, 17, and 18]. The later is characterized by its high extracting speed and no reliance on dictionary. However, it does not identify meaning (sense) of a term which is important to our IR model. For this reason, we apply a dictionary-based approach [20] to the extraction of term. Majority of the literature use patterns learned by either supervised [11] or un-supervised approaches [8, 12] to identify binary relations. But almost all these approaches are only tested on extraction of protein-protein interactions. Besides, their extracting recall is too low for IR use. We finally develop a generic ontology-based approach to extract terms and their binary relations.

Ranking is an important component to IR systems. Most existing ranking methods are directly or indirectly based on term frequency. However, frequency does not well capture the structure of terms and relations in a document. For this reason, we try to find a better metric that can fully use the information of binary relations between terms. Considering a document in our IR model can be easily formalized as a graph, $G= (V, E)$, where V denotes all terms and E denotes all binary relations, we borrow approaches and metrics from social network research [1, 6] and develop a between-ness centrality based ranking approach.

Based on the above ranking and extracting approaches, we build a prototyped IR system supporting relation-based search for Medline abstracts. We use this novel IR system to improve the retrieval result of all official runs in TREC-04 Genomics Track. The experiment shows promising performance of relation-based IR. The average P@100 (the precision of top 100 documents) for 50 topics is significantly raised from 26.37 %(the P@100 of the best run is 42.10%) to 53.69% while the MAP (mean average precision) is kept at an above-average level of 26.59%. The P@10 is also improved from 42.69% to 61.87%. The experiment also shows the expressiveness of relations for the representation of information needs, especially in the area of biomedical literature which are full of various biological relations.

The rest of the paper is organized as follows: Section 2 describes the representation of documents. Section 3 presents a generic approach to extraction of terms and relations. Section 4 shows a ranking approach for relation-based IR model. Section 5 presents the experiment design and result. A short conclusion finishes the paper.

2   Representation of Document and Query

Traditional IR models a document as a bag-of-word (a), i.e. a document consists of a set of words which are treated as independent of each other. Because a term (it is also called phrase in other work) is often more meaningful than a word, bag-of-term model (b) is naturally extended from the bag-of-word model. For example, *high blood pressure* is treated as one token instead of three tokens in bag-of-term model. A word or a term may have different meanings in different context. Thus, a bag-of-sense (c) is further evolved for information retrieval.

The above three models (a, b, and c) may produce slightly different performance for IR. But neither of them addresses the relation among tokens. Actually, a document

is often full of various explicit and implicit relations. For example, biomedical literatures contain a large number of biological interactions among gene, protein, mutation,
disease, drug, etc. Intuitively, incorporation of such knowledge (represented by relations) will help improve the performance of an IR system. For this purpose, we propose a relation-based IR model below.

Terms (CUI, String, Semantic Type, Frequency)
T1 (C0003818, arsenic, Hazardous or Poisonous Substance, 9)
T2 (C0870082, hyperkeratosis, Disease or Syndrome, 4)
T3 (C1333356, XPD, Gene, 6)
T4 (C0007114, skin cancer, Neoplastic Process, 1)
T5 (C0012899, DNA repair, Genetic Function, 3)
T6 (C0241105, hyperkeratotic skin lesion, Finding, 2)
T7 (C0936225, inorganic arsenic, Inorganic Chemical, 1)
......

Relations (First Term, Second Term, Frequency, Type)
R1 (T1, T3, 3, E)   R2 (T2, T4, 1, E)
R3 (T2, T5, 2, E)   R4 (T2, T3, 2, E)
R5 (T4, T5, 1, E)   R6 (T3, T4, 1, E)
......

Fig. 2. A real example of document representation. The document (PMID: 12749816) can be
found through PubMed. CUI is the sense ID of a concept in UMLS[3]. *E* in relation representation stands for entity-entity relation.

2.1 Document Representation

In relation-based IR model, we represent a document by a set of sense disambiguated
terms and their binary relations as shown in Figure 2. We record the sense rather than
the string as the unique identity of a term based on the following two considerations.
First, term sense can relieve the synonym problem in IR. Because all synonyms share
one sense ID, we can simply use one sense ID to find all documents containing its
synonyms without query expansion. Second, it can solve polysemous problem in IR
because a word (even a phrase) may have different meanings across documents and
queries while the sense ID never causes ambiguity [15, 19]. However, strings still
provide useful information for IR. For example, in the experiment of TREC 2004
Genomics Track (see Section 2.2), we use string to decide if a term (protein) belongs
to certain protein family. Thus, we keep the string of a term in term indices. Also, we
record the semantic type of a term, the category a term belongs to. The semantic type
is useful to express information needs (see Section 2.2).
　　A relation is defined as a pair of two terms which are semantically and syntactically related to each other. We identify all such term pairs in a document and record
their frequency. The relations fall into two types: entity-entity relation and entity-
attribute relation. The entity-entity relation addresses the interaction of two entities,

---

[3] http://www.nlm.nih.gov/research/umls/

for example, the protein-protein interaction and the relation between genes and diseases. For the simplicity, the entity-entity relation in our model is undirected. The other type of relation is entity-attribute. It is about from what point of view the entity is described. For example, in the entity-attribute relation, regulation of TGFB gene, TGFB gene is the entity and regulation is the attribute of TGFB gene. Obviously, the entity-attribute relation is directed.

2.2 Query Representation

The query representation is subject to the mechanism of document representation. Under traditional term-based IR model, we often use term vector or term-based Boolean expression to represent information needs. In this section, we will first briefly introduce the syntax of relation-based Boolean expression and then demonstrate the effectiveness of this query representation mechanism by the examples from TREC 2004 Genomics Track.

Three types of predicates, denoted by term (T), entity-entity relation (R), and entity-attribute relation (M), are available to build Boolean expression. A term can be specified by any combination of its string (STR), sense ID (CUI), and semantic type (TUI). All predicates can be combined by AND or OR operator. Here, we use the ad hoc topics in TREC 2004 Genomics Track[4] to illustrate the usage of relation-based Boolean expression to represent user information needs.

Topic #1: Ferroportin-1 in humans
Query: T (CUI=C0915115)
Notes: C0915115 is the sense ID of *Ferroportin-1* in the dictionary of UMLS (Unified Medical Language System). All term senses in this paper is based on UMLS.

Topic #2: Generating transgenic mice
Query: M (CUI_1= C0025936 AND STR_2=generation)
Notes: C0025936 is the sense ID of transgenic *mice*

Topic #12: Genes regulated by Smad4
Query: R (CUI_1=C0694891 and TUI_2=T028)
Notes: C0694891 is the sense ID of *Smad4* and T028 stands for the semantic type if *Gene*. Because entity-entity relation is undirected, the query should contain the symmetric predicate R (CUI_2=C0694891 and TUI_1=T028). However, for the simplicity, we let the IR system automatically generate the symmetric predicate R.

Topic #14: Expression or Regulation of TGFB in HNSCC cancers
Query: R (CUI_1=C1515406 and CUI_2=C1168401)
Notes: C1515406 is the sense ID of *TGFB* and C1168401 is the sense ID of *HNSCC*

Topic #30: Regulatory targets of the Nkx gene family members
Query: R (STR_1 like nkx% and TUI_1=T028 and TUI_2=T028)
Notes: we assume a term with its string beginning with nkx and with semantic type of gene is the member of *Nkx gene family*.

---

[4] http://trec.nist.gov/data/genomics/04.adhoc.topics.txt

We can see that relation-based Boolean expression is neat and powerful to express user information needs from above examples. In topic #1, we simply use one T predicate though Ferroportin-1 has lots of synonyms. In topic #12 and #30, we use one R predicate in conjunction with semantic types to express a question-answering type information need that is very difficult to be represented by term vector or term-based Boolean expression.

3 Extraction of Terms and Relations

In this section, we propose a generic ontology-based approach to extraction of terms and relations. As shown in Figure 3, we first extract terms using domain ontology in conjunction with part of speech patterns [20]; then use surrounding words to narrow down the sense; finally employ several heuristic approaches to generate relations.

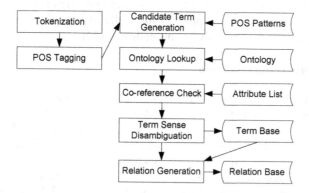

Fig. 3. The architecture of the term and relation extraction system

3.1 Extraction of Terms

There are volumes of literature on the topic of term extraction from biomedical literatures. Most of them use either hand-created rules or machine-learned rules to extract terms from text. However, neither of them extracts meaning (sense) of the term. For instance, the IE system may tell you that Ferroportin-1 is a protein but not tell you what protein it is. Because we record sense ID rather than string as the ID of a term, we use a generic ontology-based approach [20] that identify not only the category of the term, but also its possible senses. This approach begins with part of speech (POS) tagging, then generates candidate terms using POS patterns, and finally determines if it is a term by looking up the ontology.

In this particular project, we take UMLS as the domain ontology. UMLS is built from the electronic versions of many different thesauri, classifications, code sets, and lists of controlled terms in the area of biomedicine and health. The Metathesaurus of UMLS is organized by concept or meaning of terms and provides their various names (synonyms), and the relationships among them. By checking with the synonym table, we can easily determine if the candidate (generated by POS patterns listed in Table 1) is a term and retrieve possible senses if yes.

Table 1. Part of Speech Patterns and Examples. NN, NUM, and JJ denote noun, number, and adjective, respectively. All article, preposition, and conjunction words will be removed from the original text during pattern matching.

Part of Speech Pattern	Examples
NN NN NN	Cancer of Head and Neck
NN NUM NN	DO 1 Antibody
JJ NN NN	High Blood Pressure
NN NN	DNA Repair
NN NUM	Ferroportin 1
JJ NN	Sleeping Beauty
NN	FancD2

A term sometimes appears in the form of a pronoun such as *it* or its abbreviation. It is then necessary to figure out what the pronoun or abbreviation refers to in context. We develop a simple heuristic approach to handle abbreviations and implement a light method provided by Dimitrov and his colleagues [4].

3.2   Term Sense Disambiguation

A term may have multiple meanings defined in the dictionary. For example, *Ferroportin-1* has two senses defined in UMLS (*C0915115: metal transporting protein 1; C1452618: Slc40a1 protein, mouse*). Inspired by a finding that the ambiguity of many terms in text is caused by use of short name, abbreviation, or partial name, we take an unsupervised sense disambiguation approach adapted from Lesk's word sense disambiguation (WSD) approach that basically tags sense by maximizing the number of common words between the definition of candidate senses and the surrounding words of the target [10].

Different from Lesk's approach, our approach first use surrounding words (3 words in the left side of the target and 3 word in the right side of the target) to narrow down candidate senses. If there is still more than one sense left, we then score each candidate sense. In Lesk's approach, any word in any sense has same weight. Obviously it is not a good assumption for term sense disambiguation. Instead, we borrow the idea from term weighting community and use TF*IDF to score the importance of a word in a sense [9, 14]. Then the final formula to tag the sense is:

$$S = \arg\max_{j} \sum_{i} IDF_i \times TF_{ij} = \arg\max_{j} \sum_{i} \log\frac{N}{n_i} \times \frac{F_{ij}}{F_j}$$

Where:

N is the number of senses in dictionary

n_i is the number of senses containing $Word_i$

F_{ij} is the occurrence of $Word_i$ in various names of $Sense_j$

F_j is the total occurrence of words in various names of $Sense_j$

3.3 Extraction of Relations

A relation is defined as a pair of two terms which are semantically and syntactically related to each other. If there is a pre-defined relation between the semantic types of two terms in domain ontology, these two terms are simply viewed as semantically related. However, the judgment of syntactic relation between two terms is difficult. We provide two different methods of syntax judgment for entity-attribute relation and entity-entity relation, respectively.

3.3.1 Entity-Attribute Relation

If two terms within one sentence match the following pattern where *term1* is in the list of candidate attributes and the preposition is either *of* or *for*, we take *term1* as the attribute of *term2*.

Rule for entity-attribute relation: *term1* preposition *term2*
Example: *Obesity is an independent <u>risk factor</u> (term1) for <u>periodontal disease</u> (term2).*

The list of candidate attribute is compiled in a semi-automatic manner. Applying the above pattern to a sample of the collection, we obtain a list of *term1* (candidate attributes). We take the *term1* with its frequency above threshold as candidates and then have one domain expert judge its qualification for being an attribute.

3.3.2 Entity-Entity Relation

Extraction of biological interactions (relations) is a hot topic in the area of information extraction. The essence of this line of work is to generalize the syntactic form of certain relation in supervised or unsupervised manner. However, there are two major problems while applying these methods to extract biological relations for IR use. First, the indexing component of our IR model is interested in many biological relationships. But most of these reported extracting methods are tested on mere protein-protein interaction. Second, the recall of these extracting methods seems to low for IR use. For example, the IE system reported by [12] only extracts 53 relationships with 43 correct from 1,000 Medline abstracts containing the keyword "protein interaction". Instead, we employ a simple but effective heuristic approach that uses clause level co-occurrence to determine the syntactic relation of a term pair and it is able to identify various relationships with high recall and good precision for IR use.

Term co-occurrence is frequently used to determine if two terms are connected in graph-based data mining. Literature [16] takes any pair of two words in same sentence as a relation. However, as reported by [5], sentences in Medline abstracts are often very long and complex. Thus, if we follow the strategy of [16], many noisy relations may be introduced. Instead, we take clause as the boundary of a relation because terms within a clause are more cohesive than within a sentence in general. In example 1, there are three entity terms underlined and one relation (*obesity* and *periodontal disease*). The term *epidemiological study* has no relation with any of the other two terms because it is in a separate clause.

Rule for entity-entity relation: *If two terms are co-occurred within a clause, but are not coordinating components, and their semantic types are related to each other in domain ontology, this term pair is identified as an entity-entity relation.*

Example 1: *A recent epidemiological study revealed that obesity is an independent risk factor for periodontal disease.*

Example 2: *Diabetes is associated with many metabolic disorders including insulin resistance, dyslipidemia, hypertension and atherosclerosis.*

Also, Ding et al. [5] identify that coordinating is frequently occurred phenomenon in sentences and interactions (relations) between coordinating components is rare in Medline abstract. Thus, in example 2, *diabetes* has relations with remaining four terms respectively. But *insulin resistance, dyslipidemia, hypertension,* and *atherosclerosis* don't have relations with each other because they are coordinating components.

In short, we consider a term pair an entity-entity relation if these two terms are co-occurred within a clause, but are not coordinating components, and their semantic types are related to each other in domain ontology.

4   Ranking Approach

Matching the relation-based Boolean query and the relation-based representation of document, we can get a hit list for a specific query. But we still do not know the relative confidence of each document in the hit list being relevant to the query. In this section, we would answer this question, i.e. the ranking of matched documents.

A large number of term weighting schemas have been developed within TF*IDF family. The basic idea of the TF*IDF method is to synthesize the local importance of a term in a document and the global importance of a term in the collection. In general, they use inversed document frequency (IDF) to measure the global importance and use term frequency to indicate the local importance. Following this idea, we present the following framework to rank matched document:

$$R_q(d) = \sum_{p \in q} \omega_q(p) G(p) L_d(p)$$

Where $R_q(d)$ is the relevance of document d to query q, p is the predicate (term, entity-entity relation, or entity-attribute relation) that forms query, $\omega_q(p)$ is the weight of p in the query, $G(p)$ is the global importance of p in the collection, $L_d(p)$ is the local importance of p in document d.

For $\omega_q(p)$, we empirically set 0.4 for term (T), 1.0 for entity-entity relation (R), and 0.7 for entity-attribute relation (M). We still use IDF to measure the global importance of p. However, we take a metric other than *frequency* to measure the local importance. The frequency of terms or relations, of course, could be a metric of local importance because intuitively frequency is in proportion to the importance. However, frequency does not capture the structure of terms and relations in the document. That is the reason we try to find a better metric.

Since a document in our IR model is represented by a set of terms and their binary relations, it is very easy to formalize it as a graph (network), G= (V, E), where V denotes all terms and E denotes all binary relations in the document. Then we can borrow approaches and metrics in the area of social networks to measure the importance of terms (equivalent to an actor in social network) and relations (equivalent to a link in social network).

Betweenness Centrality is a frequently used metric in social network to compute the importance of an actor (a node in the network) [1, 6] and it could be extended to indicate the importance of a link (an edge in the network). The basic notion of betweenness centrality is that a vertex that can reach others on relatively short paths is relatively important. The formal definition is presented below:

In graph G= (V, E), let $\sigma_{st} = \sigma_{ts}$ denotes the number of shortest paths from $s \in V$ to $t \in V$, $\sigma_{st}(v)$ denotes the number of shortest paths from s to t where some $v \in V$ lies on and $\sigma_{st}(e)$ denotes the number of shortest paths from s to t where some $e \in E$ lies on. Then the importance of a node v is defined as:

$$C_B(v) = \sum_{s \neq t \neq v} \frac{\sigma_{st}(v)}{\sigma_{st}} \quad \text{(Freeman, 1977; Anthonisse, 1971)}$$

Similarly, the importance of a link e can be defined as:

$$C_B(e) = \sum_{s \neq t \neq v} \frac{\sigma_{st}(e)}{\sigma_{st}}$$

Normalization and edge weighting are two important issues while using betweenness centrality metrics. To control for the size of the network, both $C_B(v)$ and $C_B(e)$ will be normalized to lie between zero and one. Many social network researchers would like to normalize the betweenness centrality score by dividing the score by $(n-1)(n-2)/2$ where n is the number of nodes in the network. However, considering our purpose is to indicate the local importance of a term or relation

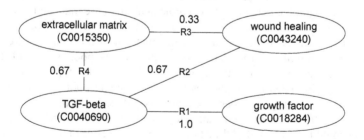

Fig. 4. A real example to calculate the local importance of relations. The document (PMID: 7929624) can be found through PubMed. The original betweenness centrality scores for relation R1, R2, R3 and R4 are 3, 2, 1 and 2, respectively. After normalization, their importance scores are 1.0, 0.67, 0.33 and 0.67.

in a document, we normalize term score and relation score by dividing their maximum value in the document, respectively. That is, the score of the most important term or relation in a document is always one.

The calculation of betweenness centrality score is related to the shortest path. Thus the weight of each edge will affect the final score. Realizing that the frequency of a relation is also an indicator to the strength or importance of the relation and related terms, we set the weight of an edge (a relation) as the inverse of the relation's occurring frequency in a document.

In short, we present a comprehensive method to rank matched documents. We consider not only the global importance of a term or a relation in the whole document collection, but also their local importance (relative importance in a document). When computing the local importance, we take into account both the structure and frequency information. The calculation of betweenness centrality score in our experiment is done by a software package JUNG[5] that implemented a fast algorithm for betweenness centrality developed by Brandes [2].

5   Experiment

In this section, we discuss the search engine and document collection used for experiment and the experiment design. Then we analyze the experiment result and compare the performance of proposed relation-based IR model with other work.

5.1   Search Engine and Collection for Experiment

To our best knowledge, there is no search engine supporting relation-based search. For this reason, we developed a prototyped IR system supporting relation-based Boolean search. We implemented conceptual document representation in Figure 2 with a DB2 database. When a query represented by relation-based Boolean expression (see Section 2.2) is submitted, the system automatically converts the Boolean expression to ANSI SQL statement and submits the SQL statement to the DB2 system.

We use the collection of TREC 2004 Genomics Track in our experiment. The document collection is a 10-year subset (1994-2003, 4.6 million documents) of the MEDLINE bibliographic database of the biomedical literature that can be searched by PubMed. Relevance judgments were done using the conventional "pooling method" whereby a fixed number of top-ranking documents from each official run were pooled and provided to an individual for relevance judgment. The pools were built from the top-precedence run from each of the 27 groups. They took the top 75 documents for each topic and eliminated the duplicates to create a single pool for each topic. The average pool size (average number of documents judged per topic) was 976, with a range of 476-1450. Based on the human relevance judgment, the performance of each official run could be evaluated (All facts and evaluation result of TREC-04 Genomics Track in Section 5 are from [7]).

Since our goal is to see whether our relation-based IR methods can further improve TREC 2004 participants' retrieval results, we build our search engine on top of search

---

[5] http://jung.sourceforge.net/

engines participated in TREC 2004. For this, we take the documents in pools for each topic and eliminate repeated documents across topics to create a single pool for our experiment use. The indexing and searching of our prototyped IR system is based on this mini-pool containing 42, 255 documents.

5.2 Experiment Design

Our goal is to build a precision-focused IR system. The major research question of this paper is *if relation-based IR outperforms term-based IR in terms of precision*. For this reason, we compare the P@100 (the precision of top 100 documents) and P@10 of our run with that of all 47 official runs participated in TREC 2004 Genomics Track. Though the overall precision (the precision of all retrieved documents) is a good proxy for precision, we do not compare this metric because TREC did not report overall precision. For convenience, we use RIR (relation-based IR) to denote our run and TREC to denote all runs in TREC 2004 Genomics Track later.

The argument of this paper that relation-based IR outperforms term-based IR in terms of precision is actually based on the assumption that explicit assertion of term relation is more useful than document level term co-occurrence to judge whether a document addresses certain relationships. To test the truth of this assumption, we study if $R (t_1, t_2)$ provides higher precision than $T (t_1)$ and $T (t_2)$ in our experiment.

We are also interested in the recall of relation-based IR though it is not our focus. On one hand, the use of relation will lower the recall because the number of documents returned by $R (t_1, t_2)$ is always equal or less than by $T (t_1)$ and $T (t_2)$. On the other hand, the use of sense instead of string well solves the synonym problem; thus it may increase the recall. So we will study the effect of use of sense and relation on the recall of IR.

5.3 Analysis of Experiment Result

Our run retrieves 125 documents on average and achieves 53.29% overall precision, 44.31% overall recall and 26.59% MAP (Mean Average Precision). MAP is a comprehensive indicator of IR performance that captures both precision and recall. As expected, the MAP of our run is at above average level. Actually it would be ranked as 15th among all 47 official runs in TREC. Our relation-based IR system can not achieve excellent MAP currently because, (1) the system is precision-focused, (2) no query expansion method is used, and (3) it uses Boolean search rather than similarity-based search. We will take (2) and (3) for future work.

The experiment shows that relation-based IR model is effective to improve the precision. We first compare the P@100 of our run with TREC runs on 50 individual topics. Except for topic 16, the P@100 of ours outperforms the average P@100 of TREC on all other 49 topics as shown in Fig. 5. The paired-sample T test (M=27.33%, t=7.413, df=49, p=0.000) shows the significant improvement of precision. Then we compare the P@100 of our run with P@100 of all official runs in TREC. As shown in Table 2, the P@100 of our run (53.69%) is significantly higher than that of the top 3 runs and the mean of all official runs (26.37%). The comparison of P@10 also supports the above conclusion. The average P@10 of TREC runs is significantly improved raised from 42.69% to 61.67%. It is worth noting that we can not say that the precision of our IR system is better than that of other IR systems because our search is based on the returns of all other IR systems. But the experiment result really tells us that the relation-based model is very promising for IR because it significantly improves the result of other IR systems.

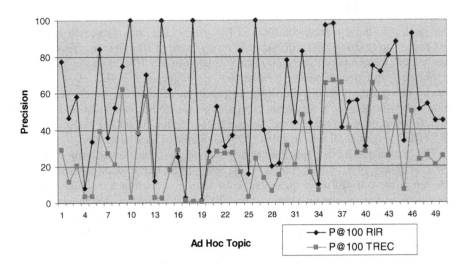

Fig. 5. The comparison of the P@100 of our run with the average P@100 of all official runs in TREC 2004 Genomic Track on 50 ad hoc retrieval topics

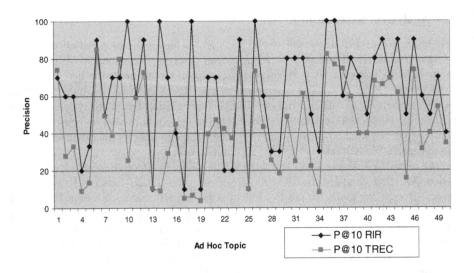

Fig. 6. The comparison of the P@10 of our run with the average P@10 of all official runs in TREC 2004 Genomic Track on 50 ad hoc retrieval topics

Table 2. The comparision of the precision of our run with official runs pariticipated in TREC 2004 Genomics Track. Runs in TREC are sorted by Mean Average Preicsion (MAP) [7].

Run	MAP	P@10	P@100
Relation IR (Our Run)	26.59	61.67	53.69
pllsgen4a2 (the best)	40.75	60.04	41.96
uwntDg04tn (the second)	38.67	62.40	42.10
pllsgen4a1 (the third)	36.89	57.00	39.36
PDTNsmp4 (median)	20.74	40.56	23.18
edinauto5 (the worst)	0.12	0.36	1.3
Mean@TREC04 (47 runs)	21.72	42.69	26.37

The rationale of relation-based IR is based on the assumption that binary relation between terms provides higher precision than document-level term co-occurrence when retrieving documents addressing certain relationships. To test the truth of this assumption, we design a simple experiment to verify. For seven queries that use a single relation (R predicate) like R $(CUI_1=A$ and $CUI_2=B)$, we change the query to the co-occurrence of two terms, i.e. T $(CUI=A)$ and T $(CUI=B)$, and search again. The experiment result is shown in Table 3. A paired-sample T test (M=11.68%, t=4.771, df=6, p=0.003) shows that the precision of relation-based query is significantly higher than that of term co-occurrence based query. This is the foundation of the argument of the whole paper that relation-based IR model contributes higher precision to domain-specific research than term-based IR models.

Table 3. The comparsion of the use of relation and term co-occurrence in IR

Topic	R (t_1, t_2)		T (t_1) and T (t_2)		P@100
	P (%)	R (%)	P (%)	R (%)	TREC04 (%)
7	35.71	8.70	24.62	27.83	27.04
8	52.00	8.07	41.05	24.22	20.94
13	12.00	12.50	8.77	20.83	2.74
14	100.00	23.81	80.00	23.81	2.70
15	61.90	14.44	48.08	27.78	18.00
21	71.43	18.75	52.83	35.00	27.96
22	29.22	46.19	25.14	65.71	27.09

Table 4. The comparsion of sense-based search and string-based search

Topic	String B	T(CUI=A)		T (STR like %B%)		T(STR=B)	
		P (%)	R (%)	P (%)	R (%)	P (%)	R (%)
1	Ferroportin	77.59	56.96	84.62	41.77	88.46	29.11
6	FancD2	84.09	39.36	84.09	39.36	85.29	30.85
9	mutY	73.38	98.26	81.75	97.39	81.48	95.65
35	WD40	97.16	63.10	99.28	50.55	98.28	21.03
36	RAB3A	98.10	81.50	98.10	81.50	98.53	79.13
43	Sleeping Beauty	80.56	14.87	77.42	12.31	77.42	12.31
46	RSK2	92.59	12.69	82.76	12.18	89.47	8.63

Table 5. The comparison of our run with runs in TREC on MAP, P@10, and P@100

Topic	Pool	DP	Hits	Rel.	MAP (%)		P@10		P@100	
					RIR	TREC	RIR	TREC	RIR	TREC
1	879	79	58	45	56.96	30.73	70.00	73.83	77.59	28.91
2	1264	101	30	14	11.88	5.79	60.00	27.87	46.67	11.66
3	1189	181	36	21	11.60	9.50	60.00	32.98	58.33	20.40
4	1170	30	167	10	56.67	2.98	20.00	8.94	8.00	3.60
5	1171	24	3	1	8.33	5.64	33.33	13.40	33.33	3.49
6	787	94	44	37	39.36	39.93	90.00	84.68	84.09	39.38
7	730	115	28	10	8.70	20.06	50.00	49.36	35.71	27.04
8	938	161	25	13	8.07	9.75	70.00	38.72	52.00	20.94
9	593	115	154	113	98.26	61.14	70.00	79.57	75.00	61.96
10	1126	4	3	3	75.00	58.11	100.00	25.32	100.00	2.77
11	742	111	215	85	76.58	32.69	60.00	58.94	38.00	38.43
12	810	256	255	174	67.58	42.25	90.00	72.32	70.00	58.66
13	1118	24	25	3	12.50	2.88	10.00	10.21	12.00	2.74
14	948	21	5	5	23.81	4.79	100.0	8.94	100.00	2.70
15	1111	90	21	13	14.44	13.88	70.00	29.15	61.90	18.00
16	1078	147	24	6	4.08	19.26	40.00	44.89	25.00	28.83
17	1150	3	66	2	66.67	8.85	10.00	5.11	3.03	1.15
18	1392	1	1	1	100.00	62.54	100.0	6.60	100.00	0.72
19	1135	1	63	1	100.00	15.94	10.00	3.62	1.59	0.62
20	814	116	154	33	26.72	14.66	70.00	39.57	28.00	22.38
21	676	80	53	28	18.75	26.71	70.00	47.02	52.83	27.96
22	1085	210	332	97	44.76	13.54	20.00	42.34	31.00	27.09
23	915	158	84	31	18.35	18.35	20.00	37.45	36.90	27.47
24	952	26	24	20	76.92	59.70	90.00	74.68	83.33	16.85
25	1142	32	38	6	18.75	3.31	10.00	10.00	15.79	3.30
26	792	47	9	9	19.15	44.01	100.0	72.98	100.00	24.11
27	755	29	60	24	82.76	26.40	60.00	43.19	40.00	13.55
28	836	13	60	12	92.31	20.31	30.00	25.32	20.00	6.43
29	756	43	42	9	20.93	13.52	30.00	18.09	21.43	15.15
30	1082	165	140	104	63.03	21.16	80.00	48.72	78.00	31.13
31	877	138	84	37	26.81	9.56	80.00	24.89	44.05	20.72
32	1107	496	386	323	65.12	18.04	80.00	60.85	83.00	47.87
33	812	64	39	17	26.56	13.96	50.00	22.34	43.59	16.47
34	778	31	159	19	61.29	6.44	30.00	8.30	10.00	6.68
35	717	271	176	171	63.10	34.81	100.0	82.13	97.00	65.28
36	676	254	211	207	81.50	48.87	100.0	76.38	98.00	67.00
37	476	149	250	110	73.83	53.45	60.00	74.26	41.00	65.64
38	1165	423	177	89	21.04	14.00	80.00	59.15	55.00	40.43
39	1350	317	204	107	33.75	9.84	70.00	39.36	56.00	26.89
40	1168	277	121	40	14.44	10.80	50.00	39.36	31.00	27.96
41	880	582	637	472	80.93	33.56	80.00	67.66	75.00	65.21
42	1005	697	95	68	9.76	15.87	90.00	65.96	71.58	57.02
43	739	195	36	29	14.87	11.85	70.00	69.15	80.56	25.53
44	1224	649	477	402	62.25	13.23	90.00	61.49	88.00	46.32
45	1139	156	95	32	20.51	2.86	50.00	15.74	33.68	7.11
46	742	197	27	25	12.69	26.30	90.00	73.62	92.59	49.81
47	1450	365	318	176	49.59	6.73	60.00	31.49	51.00	23.55
48	1121	155	202	104	67.10	17.12	50.00	40.21	54.00	25.57
49	1100	73	128	61	83.56	22.79	70.00	54.04	45.00	20.49
50	1091	302	174	72	23.84	7.31	40.00	34.47	45.00	25.34
Mean	975	165	125	70	26.59	21.72	61.67	42.69	53.69	26.37

Sense-based search can raise the recall of IR especially when a term has lots of synonyms because all synonyms share one sense ID. To test this hypothesis, we design the following small experiment. For seven single-term (T predicate) queries listed in Table 4, we compare the recall of sense-based search with string-based search. As expected, for the recall of topic 1 and 35, sense-based search is significantly higher than that of string-based search because both of them have many synonyms.

In this section, we successfully tested our major hypothesis that our relation-based IR model would outperform term-based IR models in terms of precision for domain-specific search. Furthermore, we tested the truth of the assumption of the major hypothesis, i.e. binary relation between terms would provide higher precision than term co-occurrence when retrieving documents addressing certain relationships. Last, we found that sense-based search would contribute higher recall to IR than string-based search especially when the searching term has many synonyms.

6 Conclusions and Future Work

In this paper, we proposed a novel relation-based information retrieval model for biomedical literature search. Unlike traditional term-based IR models that use term to index and search documents, our relation model uses sense disambiguated terms and their binary relations to index and search documents. We further develop a between-ness centrality based ranking approach that captures both frequency and structure of terms and relations. Because relations provide much contextual information and domain knowledge for IR, the use of relation may improve the precision of domain-specific IR. The experiment on a subset of the document collection of TREC 2004 Genomics Track successfully tested this hypothesis. Besides, we can draw another three conclusions from the experiment:

- An explicitly asserted relation in text is a stronger indicator of a document that addresses certain relationships between terms than the document level term co-occurrence.
- Sense-based search will bring higher recall than string-based search especially when a searching term has many synonyms.
- Relation-based Boolean expression is powerful and effective to express domain-specific information needs.

For future work, we will continue to refine the ranking approach. Though our focus is the precision, we still pay our attention to the comprehensive performance of the IR system. For this reason, we will try to extend the Boolean search to similarity-based search and find appropriate query expansion approach for relation-based IR model. Both of them will improve the recall of the IR system. We will also take effort on term and relation extraction that would further improve the performance of relation-based search.

References

1. Anthonisse, J. M., "The rush in a directed graph", Technical Report BN 9/71, Stichting Mathematisch Centrum, Amsterdam, 1971.
2. Brandes, U., "A faster algorithm for Betweenness centrality", *Journal of Mathematical Sociology,* 2001, 25(2), 163-177.

3. Deerwester, S., Dumais, S. T., Furnas, G. W., Landauer, T. K., and Harshman, R., "Indexing by latent semantic analysis", *Journal of the American Society for Information Science*, 1990, 41(6), pp. 391-407.
4. Dimitrov, M., Bontcheva, K., Cunningham, H., and Maynard, D., "A Light-weight Approach to Coreference Resolution for Named Entities in Text", *Proceedings of the Fourth Discourse Anaphora and Anaphor Resolution Colloquium (DAARC)*, Lisbon, 2002.
5. Ding, J., Berleant, D., Xu, J., and Fulmer, A.W., "Extracting Biochemical Interactions from MEDLINE Using a Link Grammar Parser", *In the 15th IEEE International Conference on Tools with Artificial Intelligence (ICTAI'03)*, 2003.
6. Freeman, L. C., "A set of measures of centrality based on Betweenness", *Sociometry*, 1977, 40:35-41.
7. Hersh W, et al. "TREC 2004 Genomics Track Overview", The thirteenth Text Retrieval Conference, 2004.
8. Hu, X., Yoo, I., Song, I.Y., Song, M., Han, J., and Lechner, M., "Extracting and Mining Protein-Protein Interaction Network from Biomedical Literature", *Proceedings of the 2004 IEEE Symposium on Computational Intelligence in Bioinformatics and Computational Biology*, 2004.
9. Jones, K.S., "Exhaustivity and specificity", *Journal of Documentation*, 1972, Vol. 28, pp.11-21.
10. Lesk, M., "Automatic Sense Disambiguation: How to Tell a Pine Cone from and Ice Cream Cone", *Proceedings of the SIGDOC'86 Conference, ACM*, 1986.
11. Mooney, R. J. and Bunescu, R. "Mining Knowledge from Text Using Information Extraction", *SIGKDD Explorations* (special issue on Text Mining and Natural Language Processing), 7, 1 (2005), pp. 3-10.
12. Palakal, M., Stephens, M.; Mukhopadhyay, S., Raje, R., ,Rhodes, S., "A multi-level text mining method to extract biological relationships" , *Proceedings of the IEEE Computer Society Bioinformatics Conference (CBS2002)*, 14-16 Aug. 2002 Page(s):97 - 108
13. Ponte, J.M. and Croft, W.B., "A Language Modeling Approach to Information Retrieval", Proceedings of the 21st annual international ACM SIGIR conference on Research and Development in Information Retrieval.
14. Salton, G., Wu, H., and Yu, C.T., "The measurement of term importance in automatic indexing", *Journal of the American Society for Information Science*, 1981, 32(3), pp.175-186.
15. Sanderson, M. 1994, "Word sense disambiguation and information retrieval", *Proceedings of the 17th annual international ACM SIGIR conference on Research and development in information retrieval*, p.142-151, July 03-06, 1994, Dublin, Ireland.
16. Schenker, A., Last, M., Bunke, H., and Kandel, A., "Clustering of Web Documents Using a Graph Model", *In A. Antonacopoulos & J. Hu (Eds.), Web Document Analysis: Challenges and Opportunities*, 2003.
17. Soderland, S., Fisher, D., Aseltine, J., and Lehnert, W., "CRYSTAL: Inducing a Conceptual Dictionary", *Proceedings of the Fourteenth International Joint Conference on Artificial Intelligence*, 1995, pp. 1314-1319.
18. Soderland, S., "Learning Information Extraction rules for Semi-structured and free text", *Machine Learning*, Vol. 34, 1998, pp. 233-272.
19. Stokoe, C. and Tait, J. I. 2004. Towards a Sense Based Document Representation for Information Retrieval, in *Proceedings of the Twelfth Text REtrieval Conference (TREC)*, Gaithersburg M.D.
20. Zhou, X., Han, H., Chankai, I., Prestrud, A., and Brooks, A., "Converting Semi-structured Clinical Medical Records into Information and Knowledge", *Proceeding of The International Workshop on Biomedical Data Engineering (BMDE) in conjunction with the 21st International Conference on Data Engineering (ICDE)*, Tokyo, Japan, April 5-8, 2005.

Author Index

Akutsu, Tatsuya   36
Amble, Tore   68

Chiang, I-Jen   9
Ching, Wai-Ki   36
Chuang, Jiin-Haur   97

Huang, Tian-Hsiang   83
Huang, Zhong   1
Hu, Xiaohua, 1, 112

Jin, Bo, 25

Lee, Yi-Shiun   83
Li, Guangren   112
Lin, Tsau Young ('T. Y.')   9
Lin, Xia   112
Liu, Charles C.H.   9
Li, Yun   1

Ng, Michael K.   36

Peng, Zhiyong   49

Sætre, Rune   68
Shi, Yuan   49
Søvik, Harald   68

Tsai, Ginni Hsiang-Chun   9
Tseng, Vincent S.   97
Tsuruoka, Yoshimasa   68

Wang, Hei-Chia   83
Wong, Jau-Min   9

Yu, Hui-Hsieh   97

Zhai, Boxuan   49
Zhang, Shu-Qin   36
Zhang, Xiaodan   112
Zhang, Yan-Qing   25
Zhou, Xiaohua   112

Lecture Notes in Bioinformatics

Vol. 4075: U. Leser, F. Naumann, B. Eckman (Eds.), Data Integration in the Life Sciences. XI, 298 pages. 2006.

Vol. 4070: C. Priami, X. Hu, Y. Pan, T.Y. Lin (Eds.), Transactions on Computational Systems Biology V. IX, 129 pages. 2006.

Vol. 3939: C. Priami, L. Cardelli, S. Emmott (Eds.), Transactions on Computational Systems Biology IV. VII, 141 pages. 2006.

Vol. 3916: J. Li, Q. Yang, A.-H. Tan (Eds.), Data Mining for Biomedical Applications. VIII, 155 pages. 2006.

Vol. 3909: A. Apostolico, C. Guerra, S. Istrail, P. Pevzner, M. Waterman (Eds.), Research in Computational Molecular Biology. XVII, 612 pages. 2006.

Vol. 3886: E.G. Bremer, J. Hakenberg, E.-H.(S.) Han, D. Berrar, W. Dubitzky (Eds.), Knowledge Discovery in Life Science Literature. XIV, 147 pages. 2006.

Vol. 3745: J.L. Oliveira, V. Maojo, F. Martín-Sánchez, A.S. Pereira (Eds.), Biological and Medical Data Analysis. XII, 422 pages. 2005.

Vol. 3737: C. Priami, E. Merelli, P. Gonzalez, A. Omicini (Eds.), Transactions on Computational Systems Biology III. VII, 169 pages. 2005.

Vol. 3695: M.R. Berthold, R.C. Glen, K. Diederichs, O. Kohlbacher, I. Fischer (Eds.), Computational Life Sciences. XI, 277 pages. 2005.

Vol. 3692: R. Casadio, G. Myers (Eds.), Algorithms in Bioinformatics. X, 436 pages. 2005.

Vol. 3680: C. Priami, A. Zelikovsky (Eds.), Transactions on Computational Systems Biology II. IX, 153 pages. 2005.

Vol. 3678: A. McLysaght, D.H. Huson (Eds.), Comparative Genomics. VIII, 167 pages. 2005.

Vol. 3615: B. Ludäscher, L. Raschid (Eds.), Data Integration in the Life Sciences. XII, 344 pages. 2005.

Vol. 3594: J.C. Setubal, S. Verjovski-Almeida (Eds.), Advances in Bioinformatics and Computational Biology. XIV, 258 pages. 2005.

Vol. 3500: S. Miyano, J. Mesirov, S. Kasif, S. Istrail, P. Pevzner, M. Waterman (Eds.), Research in Computational Molecular Biology. XVII, 632 pages. 2005.

Vol. 3388: J. Lagergren (Ed.), Comparative Genomics. VII, 133 pages. 2005.

Vol. 3380: C. Priami (Ed.), Transactions on Computational Systems Biology I. IX, 111 pages. 2005.

Vol. 3370: A. Konagaya, K. Satou (Eds.), Grid Computing in Life Science. X, 188 pages. 2005.

Vol. 3318: E. Eskin, C. Workman (Eds.), Regulatory Genomics. VII, 115 pages. 2005.

Vol. 3240: I. Jonassen, J. Kim (Eds.), Algorithms in Bioinformatics. IX, 476 pages. 2004.

Vol. 3082: V. Danos, V. Schachter (Eds.), Computational Methods in Systems Biology. IX, 280 pages. 2005.

Vol. 2994: E. Rahm (Ed.), Data Integration in the Life Sciences. X, 221 pages. 2004.

Vol. 2983: S. Istrail, M.S. Waterman, A. Clark (Eds.), Computational Methods for SNPs and Haplotype Inference. IX, 153 pages. 2004.

Vol. 2812: G. Benson, R.D. M. Page (Eds.), Algorithms in Bioinformatics. X, 528 pages. 2003.

Vol. 2666: C. Guerra, S. Istrail (Eds.), Mathematical Methods for Protein Structure Analysis and Design. XI, 157 pages. 2003.